WATER POLLUTION CONTROL ENGINEERING

Some British achievements

CONSULTING EDITOR: DR A. L. DOWNING,
DIRECTOR OF THE WATER POLLUTION
RESEARCH LABORATORY, STEVENAGE,
HERTFORDSHIRE

Prepared for the Board of Trade
by the Central Office of Information, London, 1969

contents

Foreword

BY ERIC FELGATE,
DIRECTOR OF PRODUCTION AND TECHNOLOGY, CONFEDERATION OF BRITISH INDUSTRY

Britain is relatively small in land area but has been highly industrialised for a long time. An inevitable consequence of this is the need for adequate supplies of water for industrial purposes and the need to dispose of industrial effluent. An additional problem is the disposal of domestic effluent from an increasing population concentrated in a limited number of areas.

The water used for public supply and industrial purposes in Britain is a significant and steadily rising proportion of the total available from natural sources. This has required the re-use of river water already containing a significant proportion of effluent, and a re-use of up to 10 times is not unknown.

To cope with the increasing need for water and the basic requirement of maintaining rivers in a relatively healthy state, a series of legislative measures has been introduced in Britain, the first important Act being the River Pollution Prevention Act 1876. This has led to the design, development and construction of plant for the treatment of effluent.

The chapters of this book describe the problems of effluent control, and the knowledge and specialised plant available to deal with them. These problems are rapidly becoming internationally significant and this book is recommended for study by all those, wherever they may be, who are faced with problems of effluent disposal and prevention of river pollution.

CHAPTER 1

Control of Water Pollution in Britain

BY A. KEY, CBE, DSc, PhD, FInstWPC

The management of water resources, including effective control of pollution, is one of the major problems facing industrial nations today. In this field, as in many other industrial developments, Britain has had a much longer experience than most other countries and has been a pioneer in scientific research regarding it.

As early as 1850, river pollution was one of the major topics of national debate in England, largely because the Thames, the river chiefly affected, passed through London and under the very walls of the Houses of Parliament. Moreover, the Thames was then the main water supply for a large and rapidly growing industrial community and received all its crude sewage and untreated wastes—with calamitous results, revealed by the appalling incidence of disease and mortality at the time. Probably no other capital city had ever found itself faced with such a nauseating, offensive and dangerous pollution problem, and the violent language of parliamentary debates reflected its gravity and urgency.

The remedy found, following that outcry, was essentially simple, though by no means cheap. It consisted in draining the sewage several miles lower down the river and in discharging it at a point where it received a much greater dilution, while extracting the water supply much farther upstream. Later, more and more advanced techniques of storage, filtration and sterilisation were used, with the result that, for many years now, London's water supply has been beyond reproach.

The main pollution problem then moved to the north of the country, where there was a large concentration of industry. There, it was more readily tolerated, because ample supplies of pure water were available in the hills surrounding the towns and because amenity was then considered of less importance than industrial

1. Opposite: the Crossness sewage treatment works of the Greater London Council

prosperity. The revealing dictum, 'where there's muck, there's money', may well have been coined at the time.

However, with the years, water pollution became progressively worse, but, as it remained short of disastrous, national concern was not effectively raised. When the first Act of Parliament to control pollution was passed in 1876, it was largely ineffective, because its provisions were not strong enough and also because no adequate administrative organisation was set up to implement it.

Nevertheless, important developments in pollution control were quietly taking place. Before the end of the nineteenth century, percolating filters were in large-scale use in Britain. In 1913, three British scientists, Fowler, Ardern and Lockett, read at a Manchester hotel their epoch-making paper announcing the discovery behind the activated-sludge process. Several large and successful works employing this process were constructed. In 1915, the Royal Commission on Sewage Disposal completed the work on which it had been engaged since 1898, and one or two prototype organisations for administering water pollution laws were created, paving the way for more recent developments.

By the end of the second world war, conditions were ripe for the establishment of modern legislation and organisations to deal with the whole problem. Looking back after a further 20 years on what was then progressively brought into being, it may now be seen that it really does fit the circumstances of a densely populated industrial state. It is probable that, with small modifications, it would also suit countries in other stages of development.

The present system of pollution control in Britain was made possible because:

(1) it was recognised that, contrary to what had been previously assumed, pollution is not merely a local problem, and that its effects can be serious at long distances from its point of origin;

(2) it was realised that the demand for water had reached such dimensions that it was no longer possible to use it once and then discard it: it had to be reclaimed for subsequent uses;

(3) public opinion had come to realise that attractive rivers supporting a normal aquatic life constituted an amenity of great value, well worth preserving or restoring.

The remainder of this chapter relates specifically to England and Wales. The situation in Scotland is different, because the population there is not so dense, rainfall is generally greater, and much industrial development is near the sea, which facilitates waste disposal.

In England and Wales, then, there are nearly 30 river authorities, covering together the whole country. Each is responsible for the catchment area of either a single river basin or that of a number of smaller neighbouring river basins. In each case, all streams, large or small, are under the charge of a single authority, which is fully representative of the interests involved, and staffed by highly qualified engineers, chemists, biologists and geologists, as may be necessary. These authorities carry out the duties assigned to them by various Acts of Parliament, under the control powers vested in them by law. Their duties include control of drainage, fisheries, water pollution and water conservation. In brief, they are responsible for the management of water resources in their areas, and as such they are answerable to the Minister of Housing and Local Government. There is also a Water Resources Board giving specialist advice and taking a long view of the development of regional and national water resources.

Britain, and particularly England, is one of the most densely populated countries in the world. It is a relatively small island, much of which is hilly and even mountainous. Rivers are therefore, on the whole, quite short and fairly swift, so that although rainfall is generally ample, its 'residence time', without artificial restriction, is short: it quickly flows to the sea. There is, however, a great deal of permeable ground in the south and east of England from which rain percolating in the winter can be pumped up to supply about one-third of the population with high-class drinking water. But for almost all the industrial north and midlands, water supply has depended hitherto on the artificial impoundment of the headwaters of the many streams rising in the hills. Both those sources of drinking water are now fully utilised, while the demand for it continues to grow. So does the demand for water for other purposes—industry, irrigation, amenity, etc. For all these, we have to look increasingly to rivers which are on the whole short and small, and which are at the same time the only places into which we can discharge effluents.

If water extraction were permitted without limit, some rivers would cease to flow in dry periods; if effluents were diverted elsewhere (which is generally impracticable), there would be insufficient water for many purposes; and if effluents were not controlled as to composition and purity, the quality of the water in the rivers would prevent their proper use: they would become poisonous to fish, and probably to man, as well as offensive. It is this situation, getting worse every year, with which river authorities have to deal.

For this, they have three legal tools. Firstly, no abstraction of water can be made without a licence from them, for which a charge can be made. Secondly, they can provide extra water, for instance by building reservoirs into which flood water can be pumped and allowed to re-enter the river when flows are low; the cost of such schemes is necessarily high, and can be covered by the abstraction charges just mentioned. Thirdly, they can control polluting discharges so as to maintain stream quality in a desirable condition. This third power is the most important in the present connection.

When this power was given to river authorities in 1951, it was believed that it would be possible to issue regulations laying down the permissible quality and quantity of effluents that could be discharged into a whole river basin, or at least substantial parts of it, and that, once this had been done, every discharger would know what was required of him and would act accordingly. The river authority would merely have to check that the limits were being conformed with, and take action if they were not.

In the succeeding 10 years, however, it was realised that this procedure was not really practicable and that, even if it were, it would not be the best, simplest, or most economic way to deal with the problem. Instead, each discharge is now considered on its merits, bearing in mind its quantity, nature and composition, the availability of purification processes, the size and type of the river involved, and the use to which the water is put. After considering all these factors, and often after detailed discussion with the prospective discharger, the river authority issues a consent to the discharger, laying down conditions as to both maximum quantity and maximum limits of impurity, as indicated by certain general or specific tests. In the event of disagreement amounting to an impasse, the Minister of Housing and Local Government is appealed to, and his decision is final.

It cannot be claimed that the situation just described is as yet fully established, or that the standards laid down are yet reached, or even that all the standards are as high as the river authorities hoped, but much progress has been made and it is now generally agreed that, in spite of the rapidly increasing production of liquid wastes, the rivers of Britain have recently definitely improved.

All this has cost a great deal of money—and much more will be needed. Because of the smallness of most rivers and the large volume of effluent, a high degree of treatment is usually called for. The volume of sewage—about 30 imperial gallons (0·136 m^3) per person per day—is not unduly large by modern standards, but the sewage is strong—BOD (biochemical oxygen demand) 350 mg/l on the average—and is produced by almost everybody, because virtually 100 per cent of the population have piped water supplied to their houses. To this can be added half as much again of industrial effluents which are mixed with municipal sewage and treated and discharged with it, as well as vast quantities of industrial effluents treated and discharged separately.

The total cost of purifying all this is not accurately known, but it is revealing to give examples of loans raised by municipalities during the last few years for sewerage and sewage treatment works in England and Wales. The figures were as follows—1957: £28 million; 1959: £39 million; 1961: £50 million; 1965: £58 million; and 1967: £72 million. There is no sign of the increase coming to an end.

The reason for this very considerable expenditure is the high degree of effluent purification required. It is now over 50 years since the Royal Commission laid down its 'standard' of 30 mg/l suspended solids and 20 mg/l BOD. This standard, which has wide application, represents a 95 per cent reduction in BOD, but in a growing number of cases it is inadequate, a reduction of 98 per cent or more being called for. Only thus can some of our heavily loaded rivers be kept in good condition. This situation is to be contrasted with that of many large cities in other parts of the world, situated on such large rivers that it is only necessary to give primary treatment to sewage, reducing its BOD by 50 per cent or less.

Two questions must be asked and answered when expenditure of the above magnitude is being contemplated: is it necessary, and is it giving value for money? These are very large questions, on which a great deal of investigation has been and is being carried out in Britain.

The largest water pollution research organisation in this country is the Ministry of Technology's Water Pollution Research Laboratory at Stevenage, Hertfordshire, which has earned a very high reputation for itself, notably in the last 20 years. Some river authorities, and the larger sewerage authorities, carry out additional research; some universities also play their part. Moreover, manufacturers of treatment plant sponsor a great deal of research and development work.

The question 'Is it necessary?' is important because, on the one hand, if too little effluent purification is carried out the whole object of the exercise fails and the money has been ill spent; on the other hand, if more purification is effected than is necessary the excess is just wasted. To do the right job in the right way thus prevents waste of money, and it is worth considerable effort to find out what it is. To this end, the Water Pollution Research Laboratory is making sustained investigations into the effects of pollutants of different kinds and in different quantities on the several aspects of river life, and they are all relevant to the question.

To the same end, river authorities make their own surveys and investigations. It is already possible to say, far more precisely than a decade or two ago, by how much particular effluents should be purified and in what direction, so as to maintain a river in the condition required by the uses made of it. Soon it may be possible to carry out accurate calculations. For a river receiving a large number of polluting effluents, there will doubtless be many ways in which the desirable quality may be achieved. These will include provision of more diluting water and diversion of some effluents to other discharge points, etc—but the crux of those calculations is that they can determine the cheapest, most efficient and most beneficial combination in a way that cannot be gainsaid.

The second question—value for money—implies the development of processes doing the right job efficiently, that is, with the least possible demand on the resources of a nation, community or individual enterprise. The Water Pollution Research Laboratory devotes a good deal of effort to this problem, as do also the British firms who develop, build and market treatment plants. For a number of years now, there has been no revolutionary development in the field of water pollution control that could be compared with, say, the invention of the jet engine in aviation, but there has been, nevertheless, steady progress all along the line. Better designs and controls, more knowledgeable management, are all playing their part in cutting the cost of treatment processes. Notable advances have been achieved particularly in three matters that are worth mentioning briefly in this opening chapter.

The first relates to synthetic detergents. The effect of biologically resistant detergents is well known. In the late 1940s and early 1950s, the mountains of foam they formed on many rivers wherever the water passed over a weir caused a public outcry. A committee was set up representing Government, industry, sewage treatment authorities and research organisations. It secured full technical co-operation in developing alternative materials which could be broken down biologically. Large-scale tests, one involving a town of 100,000 people, were carried out and, as an interim measure, about 50 per cent of the non-degradable detergents were replaced by degradable ones. Finally, an agreement was reached and, from the end of 1964—earlier than anywhere else—the original type of detergents was replaced by one that could be almost fully destroyed during normal sewage treatment. A degree of purification exceeding 90 per cent is now common, and detergent levels in effluents are but a small fraction of what they were. All this was done without compulsion, without subsidy and without legislation—just by co-operation and public spirit.

The second matter concerns the river Lee, which runs southward to join the Thames as it leaves London. It is a smallish river, but it forms an important source of the capital's water supply. At the end of the second world war, its quality was, to say the least, not good, and was a matter of considerable concern to the water supply authority. Since then three new towns, Harlow, Stevenage and Hatfield, have been built; others, including Luton, have been expanded considerably; and much development has taken place along the Lee valley—all discharging their wastes to the river. However, such is the efficiency now achieved by the treatment of sewage and industrial effluents in the area that the river Lee is in a far better condition than it was with the lighter population load. Much of this improvement has been due to the development of tertiary treatment processes, about which more will be said in subsequent chapters. Some of the major sewage effluents to the river now have BODs averaging as low as 5 mg/l for long periods, and 99 per cent of the ammonia contained in the sewage is oxidised before discharge.

The third 'success story' concerns the tidal part of the river Thames. In spite of the early action described above, its waters had deteriorated again until, shortly after the last war, stretches of it became deoxygenated and offensive. This was due to a combination of causes, most of which could be removed or rendered less effective—at a price. Which was the best way to go about it? The answer was provided by a detailed investigation, extended over some 15 years. A mathematical model of the actual situation was devised, from which the effect of any specified altered circumstance could be predicted. Tests of the model indicated that its prediction accuracy could provide a fully adequate basis for future management of the waters in the estuary, and it is being used for that purpose. Already, many of the indicated remedial works have been put into operation, and the river is no longer offensive.

The case of the Thames estuary, though difficult enough, is simple in two respects: there is no substantial stratification caused by the fresh water flowing above the sea water; and, for the present, a single criterion (dissolved oxygen) is sufficient to assess water quality. Other British estuaries are more complicated, but some of them are being studied and the prospects are that the behaviour of these, too, will be reduced to predictable and thus remediable dimensions.

A. KEY

After graduating in chemistry at Leeds University, Dr Key spent several years carrying out research for the gas industry on the gasification of coke and later transferred his attention to the treatment of gas liquor. This necessitated a knowledge of sewage treatment and river science and paved the way for his appointment in 1944 as chemical inspector to the Ministry of Health (later to the Ministry of Housing and Local Government). For many years he was senior chemical inspector, and is now consultant. Dr Key became internationally known when he prepared a detailed account of Water Pollution Control in Europe for the World Health Organisation. He has served on a number of expert committees of both the WHO and the Economic Commission for Europe. In 1964 he gave the keynote address to the International Conference on Water Pollution Research in Tokyo. He has been chairman of British committees on toxic solid waste disposal, sewage and water analysis, and coastal pollution research.

CHAPTER 2

Scientific Management of Pollution Control

BY B. A. SOUTHGATE, CBE, BA, PhD, DSc, FRIC

Since the first major Act of Parliament to control water pollution, passed in 1876, British experience has been that it is not sufficient just to prohibit the discharge of polluting matter. For the system to be successful, it is necessary to define what 'polluting matter' is and to formulate exactly the limits to the properties and concentration of named constituents in effluents that may not be exceeded. Whether an effluent complies with the standards set should be readily ascertainable, both by those who discharge it and those who impose the standards, using officially recognised methods of chemical, physical or biological examination.

Effluents differ so much in their volume, composition and properties, while rivers vary so much in their size and in the use made of them, that control standards must inevitably vary in kind and severity for different effluents, if the quality of surface waters is to be maintained at an acceptable level without imposing unnecessary expense on local authorities and industry. Scientific management of the water economy requires that river authorities should be able to predict what the effect of discharging a given effluent in a specific receiving water will be. Only thus can a standard be achieved that will be adequate but not unnecessarily severe.

The importance of this type of quantitative control was appreciated in Britain more than 50 years ago by the Royal Commission on Sewage Disposal. Sitting from 1898 to 1915, it did much to develop practicable methods of treating sewage and trade wastes, and to evaluate the effects of effluents on surface waters. From visual, chemical and biological observations, it classified rivers in five grades of quality, from 'very clean' to 'bad'. It considered that, to be graded between 'fairly clean' and 'doubtful', river water should contain not less than 60 per cent of the saturation value of dissolved oxygen in the summer and that, for this to be so, its biochemical oxygen demand (the test for which was devised by the Commission) should not exceed 4 mg/l. From this it was concluded that the BOD of a sewage effluent should not normally exceed 20 mg/l, since with little dilution in a river—the then generally available minimum dilution ratio of 8 : 1 was taken as normal—the effluent would convert a clean river (BOD 2 mg/l) to no worse than a 'doubtful' one.

The test for biochemical oxygen demand introduced by the Commission is still by far the most widely used to give an estimate of the oxygen-absorbing capacity of effluents. Moreover, the limiting value of 20 mg/l for sewage effluents and organic industrial wastes of similar character, though it never had general legal force in Britain, has been widely accepted by river authorities as appropriate for liquids discharged to freshwater rivers, unless there be special reasons why a more stringent standard should be necessary or a more lenient one allowable.

The Commission thus laid down a sound foundation on which to base a system of pollution control. Since it published its Report in 1915, however, great changes have occurred. The volume of water used has steadily increased, and with it the proportion of industrial and sewage effluents in rivers from which many lowland water supplies have to be taken, or will have to be in the future. There has been a widespread public demand that long-standing pollution of certain rivers and estuaries in industrial areas should be abated, so that these waters can again be used for fishing or recreation. At the same time, the task has been made more difficult by a great increase in the volume of industrial effluents, and particularly in the growing variety and noxious character of the constituents some of them contain.

For many years, therefore, a concerted national effort has been made by government, local authorities and industry to preserve the country's water resources. In the field of pollution control, where much of the considerable capital expenditure involved is on works which, once built, cannot easily be modified, it is important that their design should be based on an accurate knowledge of what it is intended to achieve and of the ability of the plant to achieve it.

Much research has been necessary. Because of the manifold effects of polluting substances, and in view of the very large volume of sewage and industrial effluents to be dealt with, this has involved work in many disciplines, notably in biology, chemistry, mathematics, medicine, physics and engineering. The economic solution of some industrial problems may lie in the recovery and marketing of a saleable product from what would otherwise be a polluting waste water.

In general, the two chief aims must be (1) to determine precisely the effects of specific waste products, or classes of products, on the suitability of surface waters for the uses to be made of them; and (2) to devise practicable and economic methods of reducing the quantities of the discharged substances to a level at which they do not preclude the use of water for the purposes required.

An example of the complexity of the research which may be necessary where very large expenditure is involved is afforded by an investigation of the Thames estuary which was completed in 1964. This waterway, serving London, the country's busiest port, receives the domestic and industrial effluents from a population of over 6,000,000 people. More than a century ago, sewers were built to carry sewage from the central part of London to outlets on both banks, in the middle reaches of the estuary. As London grew, the flow increased—it is now some 300 million imperial gallons (1·36 million m^3) per day—and, though partial treatment plants were provided, the middle reaches of the estuary had become, by 1948, devoid of dissolved oxygen during dry weather, the sulphate of the sea water was being reduced by bacteria to sulphide under the anaerobic conditions, and a nuisance had developed.

Before embarking on the large expenditure that was obviously required to put things right, answers to several questions had to be found. The temperature of the water in the most polluted reaches had risen by about 5°C since 1900, as a result of heat discharged from electricity generating stations. Would the provision of cooling towers make an appreciable difference to the condition of the estuary? Abstraction of water from the upper Thames, just above the head of the estuary, had doubled between 1900 and 1948, when it was about 240 mgd (1·1 million m^3/d). If greater reservoir capacity were provided and abstraction reduced during dry weather, would this effect a worth-while improvement? If treatment of sewage had to be extended, what degree of improvement would be necessary, and which discharges would it be most profitable to improve? Finally, after the estuary had been brought back to a satisfactory condition, what standards of quality would have to be imposed on any new discharges to ensure that it remained so?

To answer these questions required some 15 years of research, and involved the development of a mathematical model of the dispersal of polluting matter by tidal action, as well as a study of the oxidation rates of polluting wastes constituents and of the rate of oxygen supply from the air and from the bacterial decomposition of nitrates and sulphates. The quantities of polluting matter discharged during past years were known approximately, and the condition of the estuary water had been surveyed by the local authority since 1882. Thus, the state of the water predicted for past years could be compared with what had been observed.

Prediction proved to be satisfactorily close to observation, and the method could thus be used with confidence to forecast what the effect would be of making further changes in conditions affecting the sanitary condition—and especially the oxygen balance—of the estuary water. Eventually, it was computed that neither restriction of heat input nor of abstraction of fresh water from the upper river would be economically sound, and that the best return for money spent would come from extending the two main sewage works of London, discharging to the central reaches. Most of this extension has since been made—at a cost that will ultimately be around £40 million; the estuary is no longer anaerobic, and its oxygen content is gratifyingly close to what had been predicted.

That investigation was necessarily a lengthy one, since no study of a similar complexity and precision had been undertaken before. Now that the methods of calculation used there have proved their validity, they can be applied—often in much simpler form—to the study of other estuaries. For instance, a short investigation of the Humber gave the river authorities concerned a useful picture of the relative effects on it of different polluting discharges. Recently, a survey of the Tees estuary, made jointly by consulting engineers and the Government's Water Pollution Research Laboratory, has served as a basis for a scheme of disposal of sewage and trade wastes from the district. A great deal of similar research is in progress to elucidate the mechanism of dispersal and decomposition of sewage in the open sea, so that marine outfalls can be designed economically, yet with the assurance that neighbouring beaches will not be polluted.

It is perhaps surprising that it should have proved easier to predict the effect of polluting discharges on the level of dissolved oxygen in an estuary, with its complicated, oscillating current system, than in freshwater rivers, with their simpler, unidirectional flow. The pioneering work carried out in the United States more than 40 years ago, which led to the well-known method of predicting an 'oxygen sag-curve' below a discharge to a river, is not usually applicable to British conditions, where most rivers are comparatively shallow. For such rivers, the level of dissolved oxygen is greatly affected by organic matter deposited from an effluent onto the river bed, as well as by the photosynthesis and respiration of aquatic plants.

Though much research has been done to unravel these complications, their effect must remain in part unpredictable. Thus, when the density of plants in a river is known, their oxygen consumption by respiration can be estimated, as can also their net output of oxygen during daytime photosynthesis—provided that the light intensity is predictable, which it rarely is in Britain. However, at least some forecast can be made of the effect of plants in lowering the oxygen level during spells of bad weather, especially at night, when oxygen deficiency is likely to be of greatest danger to a fishery.

Much more accurate estimates can be made of the rate of absorption of oxygen from the air by a river during flow through its natural channel, given its depth and velocity, or, when re-aeration occurs by passage over a weir, given the height and design of the weir. The effect of insoluble organic matter on oxygen balance must remain to some extent unpredictable, since its deposition and erosion depend on sporadic changes in current speed; but, except in rapidly flowing or very deep rivers, it is usually the most important cause of de-oxygenation. As will be mentioned, special measures are taken in Britain to remove as much of it as possible from effluents discharged to streams of special importance.

In Britain, the preservation of fisheries, whether they be of economic value, as are salmon fisheries, or of interest mainly in providing recreation, as are coarse fisheries, is considered to be an important matter, and the study of the effects of pollution on fish has always formed a large part of the national research programme. Some rivers and tributaries in heavily industrialised areas contain directly toxic substances, almost always of industrial origin, and sometimes in lethal concentrations. It is in the quantitative knowledge of these, both singly and in admixtures, that the greatest progress has been made. This line of inquiry has been followed deliberately, since many of the poisons in British rivers—as, for instance, the toxic constituents of electro-plating effluents—can be removed only in plants specially designed for the purpose. As a result of pollution control by river authorities, fish have, in recent years, returned to some rivers where before they could not live. It is expected that this process will continue; but, to bring it about, the precise reasons that had made these rivers uninhabitable to fish had first to be known.

The presence of increasing proportions of sewage effluents in rivers from which water is taken for domestic supply has given rise to much research in Britain. One result of this has been the development of

2/3. Automatic control of dissolved oxygen in aeration tanks at the Rye Meads works of the Middle Lee Regional Drainage Authority: above, tanks showing position of control electrodes (E) and valves (V); below, recording and control unit

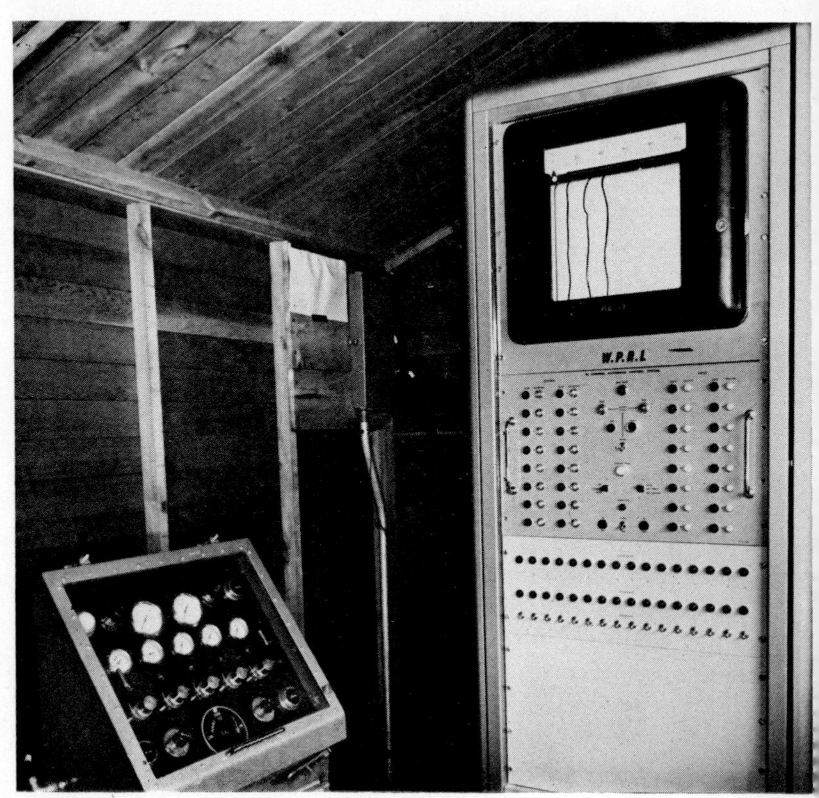

various forms of tertiary treatment of sewage effluents, designed to remove as much suspended matter as possible and thus to avoid the deposition and periodic erosion of organic sludge in the receiving stream. Another important objective has been to reduce as far as possible the concentration of ammonia in effluents, especially from the activated-sludge process, forming part of a raw-water supply, since ammonia causes difficulties when the water is treated by chlorination, the usual method of disinfection. This work was based on a fundamental examination of the kinetics of bacterial activity. Apart from defining the conditions in which removal of ammonia—by oxidation to nitrate—is possible, it has led to a much better understanding of the complex reactions occurring during the aerobic stage of sewage treatment.

One essential requirement for the oxidation of ammonia is that the concentration of dissolved oxygen in an aeration tank must not fall below a certain critical level; on the other hand, maintaining too high a concentration is wasteful. One solution is to control the supply of air automatically—a method now being tested on a large scale at a sewage works in the London area (photographs 2 and 3). The plant is one of those discharging to a river (the Lee) from which a water supply is taken. During the last full year of operation, the average BOD of its effluent was 3·6 mg/l and the average concentration of ammonia only 1 mg/l. An effluent of this quality would be suitable, after chlorination, for many industrial purposes. There is widespread interest in the possibility that, with the growing shortage of water in some parts of the country, supplies prepared from sewage effluents—especially those discharged to saline waters—will be distributed through separate mains for industrial uses for which water of the first quality is unnecessary.

A feature of the British system of pollution control is that manufacturers have a right to discharge industrial effluents to the sewers of local authorities—subject to certain conditions, of which one is that nothing discharged shall interfere unduly with the treatment of the sewage to produce an acceptable effluent. Many industrial waste products do interfere, either by inhibiting bacterial activity, or by passing unchanged through a plant and causing the effluent to be toxic or otherwise objectionable. A considerable part of the national research effort is therefore concerned with studying the precise effects of specific constituents of industrial effluents, the variety of which increases rapidly as new manufacturing processes are introduced.

Based on these findings, a local authority then issues 'Conditions of Consent' to the discharge of the effluent, specifying the properties or composition to which it must conform, if necessary after pre-treatment by the manufacturer.

It was inevitable that, in a country as industrialised and as densely populated as Britain, a strict, scientifically managed system of control would be necessary if water resources were to be preserved. Inevitably also, as demands for water increase, control will become even more stringent and precise. Already the quality of the water in some rivers is being continuously monitored by automatic instruments. The next stage is likely to be the more general adoption of this system, with communication of the results by landline or radio to river authorities' headquarters.

B. A. SOUTHGATE

Dr Southgate graduated at Cambridge in 1926, and then worked on the nutrition of farm animals. He later joined the staff of the Marine Biological Association and investigated the causes of the death of salmon in the polluted Tees estuary. Transferring to the Water Pollution Research Laboratory, he was officer-in-charge of the survey of the Mersey estuary, investigating the effects of sewage on the deposition of mud and the silting of navigable channels. In 1943 he was appointed director of water pollution research. He retired in 1966 and is now working as a consultant. Dr Southgate is an honorary member of the Water Pollution Control Federation of the USA and of the International Association of Water Pollution Research. He has been president of the Institute of Sewage Purification, and the Society for Water Treatment and Examination, and is author of Treatment and Disposal of Industrial Waste Waters.

CHAPTER 3

Sewerage

BY H. R. OAKLEY, MSc, CEng, FICE, AMIWE, MInstWPC, FASCE

The water carriage system is the basis of effective public health control in urban areas enjoying piped water supplies, and the cost of the sewerage system is frequently greater than that of the sewage treatment works. The provision of effective, economic and durable drainage systems calls for expert knowledge and wide experience. Because of their long experience in many parts of the world, British engineers are particularly well qualified to carry out work of this kind.

Sewerage systems are described with reference to the amount of rainwater run-off taken into the foul sewers. In some systems, referred to as 'separate', all such run-off from paved areas is carried by a separate system of surface water drains, and the foul sewers carry only dirty waste waters. The other extreme is the 'combined' system, in which all surface water run-off and foul sewage is conveyed in a common drainage system. Intermediate systems are usually referred to as 'partially separate', and describe situations where only part of the surface water run-off (often that from roofs and backyards of dwelling places) is taken into the foul sewers, the run-off from roads and similar paved areas draining to separate surface water sewers.

In the latter part of the nineteenth century in Britain, sewers were mostly developed as combined systems of drainage, and foul waste waters were frequently drained into existing surface water systems. For many years thereafter, combined drainage was the general rule, but recently there has been an increasing preference for partially and fully separate systems. The choice frequently depends on estimates of the relative costs of the alternative systems, but other considerations are important, including the quality of surface water discharges, pollution from storm sewage overflows on combined sewers, control of grit and other deposits in the sewers, and the costs of sewage treatment. It cannot be said that any one system is preferable in all circumstances; in most parts of urbanised Britain the separate system of sewerage is both cheaper and the more desirable, but in less developed areas, or in locations with a different rainfall pattern, partially separate or combined systems may be more appropriate.

The rate of flow in a sewer is not uniform, and even in dry weather diurnal variations occur, which are determined by living habits and industrial activity. The average daily flow of foul sewage is closely related to the water consumption, but the peak rate of flow in dry weather during the day varies according to the type and size of area served, and is frequently between two and four times the average rate over 24 hours. Also important is the minimum rate of flow, which may be as low as half the average. Because of the need to provide capacity for surface water flow, the size of sewers in separate and combined systems is largely determined by the intensity and duration of rainfall during heavy storms.

To save cost, storm sewage overflows are often provided on combined sewers. These allow relief of the sewerage system during heavy rainfall by diverting excessive flows to a stream or river, so that the sewer downstream can be made smaller. Big combined systems may incorporate tens or even hundreds of storm sewage overflows.

The overflow from combined sewers is of foul sewage more or less diluted with surface water, and is therefore polluting; indiscriminate overflow of storm sewage can result in very offensive, and perhaps dangerous, conditions. The acceptability of storm sewage overflows will depend on the frequency and strength of overflow, and on the size and nature of the receiving river, and its uses downstream.

It used to be customary in Britain to allow overflows on combined sewers when the rate of flow exceeded six times the average flow in dry weather of foul sewage, but it has long been recognised that this basis is not always satisfactory, and more stringent conditions have been set in some areas. In 1955 a Technical Committee of the Ministry of Housing and Local Government was set up to consider the whole question of storm sewage overflows and the disposal of storm sewage. An interim report was published in 1963 and the final report is now nearing completion. The committee commissioned a great deal of experimental work on the frequency and volume of overflow, the nature of the storm sewage at overflow, and the design characteristics of various types of overflow in common use, which has facilitated the rational design of combined or partially separate systems of sewers and of storm sewage overflow structures.

The run-off of rainwater into surface water sewers or combined and partially separate sewers is a function of the size, shape and slope of the area drained, the permeability of the surface, and the intensity and duration of rainfall. Whereas the physical characteristics of the drainage area are readily ascertained, rainfall is less determinate.

The maximum intensity of rainfall depends on the area covered, the duration, and the frequency of occurrence. Statistical evaluation of long-term observations is desirable, and in Britain the Meteorological Office has determined empirical formulae which can be used in surface water sewer design. The most useful is the modified Bilham formula, which relates intensity and duration of exceptional rainfall to frequency of occurrence, but it is unwise to use formulae of this type without careful examination to determine the rainfall characteristics of the particular locality.

In the design of surface water sewers, account must be taken of the physical characteristics of the area, and of the sewerage system, since the length, size and gradient of sewers determines the relationship between time and area drained, which in turn influences the likely maximum rainfall intensity and culminative run-off. The storage capacity in the sewers may also be of importance, in balancing the peak rates of flow. Improvements in design methods have been introduced by the Ministry of Transport's Road Research Laboratory, and flexible design systems suitable for computer calculation are now available for the most complicated of sewerage systems; they allow precise calculation and economic design within the limits of the basic premises.

The correct evaluation of sewer size requires a full knowledge of the characteristics of channel flow in uniform and non-uniform conditions. The normal flow range is frequently in the transitional phase between streamline and fully turbulent condition, and design has been greatly assisted by the hydraulic charts based on the Colebrook-White formula developed by the Ministry of Technology's Hydraulics Research Station.

It is important in foul sewers to ensure that solid deposits do not accumulate on the invert of the sewer, and for this reason the velocity should exceed 2 ft/sec (0·6 m/sec) at the peak daily flow rate. In normal circumstances, it is good practice for various reasons to limit the maximum velocity in the sewer to about 13 ft/sec (4 m/sec), but recent research sponsored by the Construction Industry Research and Information Association has thrown new light on the erosion of sewers due to excessive flow velocities which should be considered when gradients are steep. Consideration must also be given to the depth of flow, to prevent stranding of large solids.

In hot countries, and where the sewage is unusually strong, it may be desirable to design the sewerage system so as to limit the period of retention in the sewer, and maintain sufficient turbulence to prevent anaerobic conditions developing; but if the sewage is septic, turbulence is best avoided since it promotes the release of hydrogen sulphide gas. Adequate ventilation is always essential, for safety and amenity reasons, and where industrial wastes are present, or the sewage is septic, close attention must be given to the material of construction, or to the protection of vulnerable pipes from corrosion.

Associated hydraulic problems arise in back-drop manholes, syphons, inverted syphons and similar structures. While normal hydraulic principles can be applied, great care has to be taken with the detailed design to ensure that the sewer and appurtenant structures are self-cleansing and free from danger of blockage, accumulation of gas pockets and corrosion, and excessive noise.

Improvements continue to be made in traditional materials for sewer construction, such as glazed vitrified stoneware and pre-cast concrete.

Vitrified stoneware pipes have been improved in uniformity of material, straightness and accuracy in length and diameter, increased length of pipe, and uniformity and durability of glazing. Flexible-type joints are now commonly used both for stoneware and precast concrete pipes, and the pipes are now designed to specified strengths, to suit the more sophisticated structural design methods now used. Cast iron has been largely replaced in the smaller sizes of pipe by spun and ductile iron, which give lighter and stronger pipes more resistant to corrosion and shock loading.

Attention has also been paid to the protection of pipe materials from corrosion or chemical attack, and various types of protective coatings can be applied where necessary, while the extended use of cathodic protection of steel and iron widens the choice of pipe and facilitates economy in design.

Recently introduced materials include pitch fibre, PVC, asbestos cement and epoxy resin reinforced with glass fibre or other fabric, and, in some circumstances, these may have advantages of cost, durability and strength over traditional materials.

Concurrently with the introduction of new and improved materials, design methods have been developed which aim at economy of materials by treating the buried pipe as a structure designed to carry the specific loading applied by the ground and by vehicles or other surcharge; the depth and size of sewer, the soil condition, and the method of construction are all important factors in determining the ground pressures which result.

The present position has been summarised by an interim report of a Technical Committee of the Ministry of Housing and Local Government, which has shown by reference to experimental work in Britain that the adoption of design criteria evolved elsewhere leads to unnecessarily large factors of safety in structural design.

Competent design of a sewerage system will aim at short and shallow sewers, taking advantage of the natural fall of ground to allow as much gravity drainage as possible, but from time to time, particularly where the ground is flat, it may be necessary to lift the sewage to a higher level.

The design of sewage pumping stations follows many of the normal pumping principles, but must have regard to the variations in flow of the sewage, such characteristics as the solids content and putrescibility, possible corrosive effects, and the safety of men working in confined spaces. The wet well should be as small as possible, and variable-speed pumping with sophisticated automatic control systems is usually provided on large stations to match delivery to inflow. It is sometimes desirable to provide coarse screens at entry to remove large solids which might otherwise block the pumps, but it is always prudent to provide pumps which will allow the passage of solids of the maximum size likely to pass through house drains. There is a preference for slow-running pumps, as reliability and durability are, in many circumstances, more important than efficiency, and pumps are designed to resist wear and permit easy maintenance.

Sewage rising mains should be designed to avoid gas locking and to ensure flow velocities sufficient to prevent the accumulation of solids in the pumping system; suitable minima are 2–7 ft/sec (0·6–2·1 m/sec), according to pipe size. Surge or water hammer problems frequently arise on large and long rising mains, but can be overcome by a variety of methods which include providing additional inertia on the pumping sets, provision of air receptacles on the rising main, or modification of the delivery conditions.

Photograph 4 shows a typical large sewage pumping station. In this case, the vertical spindle pumps are direct-driven by dual-fuel engines, which normally run on sludge gas and are provided with automatic speed control regulated by the sump level so as to match the pump delivery to the incoming sewage flow.

Good sewerage design takes account of the difficulties of construction, and appraises the overall economics of the entire scheme from

4. A 4 m³/sec sewage pumping installation at Derby, England

5. *A pre-cast concrete pipe sewer laid in battered trench excavation*

he beginning of construction and throughout the useful life of the completed sewer. It is desirable that the system design be evolved with a thorough knowledge of ground and surface conditions and with the background of wide experience in constructional methods, so that the most economical arrangement can be devised.

Construction techniques vary according to the facilities available, and may affect the economic structure, so that consideration must be given at the design stage to the availability of machines and men. Thus, photograph 5 shows battered trench excavation, which may be appropriate when the ground is clear and machine excavation is economic, but requires a stronger pipeline than a sewer built in a vertical-sided trench, as shown in photograph 6, which is necessary

6. A sheet steel piled trench for a concrete sewer on a reinforced concrete bed

where the ground is unsuitable for open excavation, or the space restricted, or where hand excavation is economic. The control of ground water in trench excavation is frequently the key to ease of construction, and various methods are available, including the use of well-point dewatering sets.

In most ground and surface conditions, open trench excavation is economic to a depth of about 20 feet (6 metres), but if it is necessary to lay sewers much deeper than this, then heading or tunnel construction will be necessary. With good engineering guidance, even unskilled labour can be taught to construct difficult headings safely and expeditiously; photograph 7 shows such construction in Singapore. For larger sewers, various methods of tunnel excavation are desirable and a tunnel shield (first developed for use in London in the nineteenth century) is frequently an asset when dealing with difficult soft ground conditions; photograph 8 shows this type of construction. If it is necessary to control ground water ingress in tunnels, it may be desirable to carry out the work in compressed air; alternatively, ground freezing or chemical consolidation may be applied.

A recent technique of considerable value in passing under roads, railways or rivers is that of thrust boring or pipe pushing.

Construction of a 72-inch (183 cm) diameter sewer in heading at Singapore

In Britain, as in many other European countries, many of th main sewerage systems date from the late nineteenth century an many sewers in use today are 80 to 90 years old. Some sewers usin glazed stoneware, concrete pipes or brickwork in cement morta have proved very durable and are still satisfactory, but there ar others, such as those built with pipes having clay joints or brickwor with lime mortar, which are in poor condition.

It is sometimes cheaper to rehabilitate old sewers and to provid for their continuing use for a further period than to replace them and various methods of sewer lining have been evolved. In som instances, PVC pipes have been threaded through small sewers fo this purpose; in large sewers, a thin glass fibre or precast concret lining can be built internally, and backed with cement grout.

A particular facet of sewerage work calling for expert knowledg and technique is the design and construction of marine outfalls, whic have developed rapidly over the past decade. Outfall design requir an assessment of the behaviour of the discharge and its effect on th marine environment, but has also to take account of the construc tional aspects and the stability of line after construction.

New possibilities arise from the development of constructio methods which allow pipelines several miles long to be towed out sea or lowered from lay barges into a prepared trench on the se bed; alternatively, sections may be floated out, lowered and jointe under water; photograph 9 shows a typical industrial outfall und construction. Pipelines may be of protected steel, aluminium, PV or epoxy resin, and will need to be secured against movement on th sea bed and against floatation.

H. R. OAKLEY

Mr Oakley graduated at University College London, where he was later charge of the development of the Public Health Engineering Research Grou As a partner in the firm of J. D. & D. M. Watson, consulting engineers, has been responsible for a variety of projects on sewerage and sewage dispos including schemes for the cities of Durham, Leicester and Wakefield, the coun boroughs of Derby and Luton, and the new town of Milton Keynes. He was charge of the investigation into the marine disposal of sewage from Tyneside a the Teesside sewerage and sewage disposal study. He has also been responsi for the firm's work in Singapore and Brunei, as well as for investigations Malacca and Hong Kong.

8. Construction of a tunnel for one of the London sewers; the lining is of bolted pre-cast concrete segments

). A 9840-feet (3000 m) aluminium effluent pipeline for launching into the Severn estuary

CHAPTER 4

Sewage Treatment

BY D. H. A. PRICE, BSc, FRIC, FICE, FIPHE, AMIChemE, FInstWPC

The nature and composition of domestic sewage is basically the same for all communities of comparable social habits. In Britain the daily pollution load per head of population is represented by about 0·12 lb (55 g) BOD and the solid matter by about 0·14 lb (62 g). The strength of the sewage will depend upon the volume of water carrying this load; the water will be that used by the householder and derived from the public supply, plus any infiltration water reaching the sewers, and of course surface water where the sewerage system is not entirely separate.

The daily dry weather flow of domestic sewage per head varies between about 25 gallons (0·114 m^3) with a BOD of about 500 mg/l, and about 50 gallons (0·228 m^3) giving a sewage with a BOD of about 250 mg/l. The suspended solids concentrations are usually somewhat higher than the BOD figures, but a proportion of the sludge settling in the sedimentation tanks is derived from larger solids not included in the customary sampling; determination of suspended solids and an estimate of the sludge production based upon the determined suspended solids content and the flow will therefore be too low. Furthermore, where normal biological treatment is given secondary sludge is produced, and the total sludge to be dealt with is about 0·18 lb (82 g) per head per day. Tertiary treatment to reduce the suspended solids of an effluent to less than the usual 30 mg/l may increase the sludge by some 5 to 10 per cent.

Other components of domestic sewage, such as the free and saline ammonia and organic nitrogen, are normally correlated with the BOD and are of less importance in the consideration of design. The total dissolved solids and the chlorides will also vary with the concentration of these substances in the water supply.

Changes in domestic practices bring about alterations in the composition of domestic sewage. Anionic synthetic detergents introduced around 1950, are now present in domestic sewage in

Britain in quantities of about 0·006 lb (2·7 g) per head per day, corresponding to about 17 mg/l where the water consumption is about 35 gallons (0·16 m^3) per head per day. The increase in anionic detergents has been accompanied by a rise in the phosphate content of sewage from the polyphosphate additives in the detergent powder. Recent increases in the proportion of perborates in detergent powders have also been reflected in the boron content of sewage and treated effluent.

Kitchen waste disposal units have not so far been extensively installed in Britain, but if they were to become more generally popular there would be a substantial increase in the quantity of solid matter to be dealt with at the sewage treatment plant.

In some parts of the world, the pollution loads per head of population are materially affected by traditional foods. A vegetarian diet is likely to increase the amount of sludge to be dealt with, while one predominantly based on animal protein will tend to decrease it.

The number of treatment plants dealing with domestic sewage only is now quite small, and many sewages contain trade wastes of various types. Trade wastes may increase the BOD of the sewage, or the suspended solids, or they may introduce toxic materials which interfere with treatment. Where trade wastes are discharged to the sewer, even if they are controlled, it is desirable, if not essential, to monitor regularly the composition of the sewage arriving at the works, for there is always the possibility of illegal discharges occurring. When designs for the extension of a sewage works are in progress, it is essential to carry out sampling and analysis of the crude sewage, even if it is believed to be of domestic origin only. Such an investigation not infrequently reveals the presence of hitherto unsuspected industrial wastes.

For this purpose, single, isolated samples of the crude sewage are useless. The only satisfactory method is to sample it at regular intervals, at least hourly, but preferably every 30 minutes or even more frequently. The individual samples should be kept separate for inspection and then mixed in proportion to the rate of flow at the time of sampling. Where the flow is not known, mixing the individual samples in equal proportions will not usually introduce a serious error. Manual sampling is expensive and not always reliable. There are now available a number of well-tried automatic sampling machines, some models being battery operated and portable (as shown in photographs 10 and 11). This type of equipment is now in general use in Britain for sampling the sewage at different stages

10. An automatic sampling machine in which anything tha passes the inlet to the pump is recorded in one of the 48 bottle

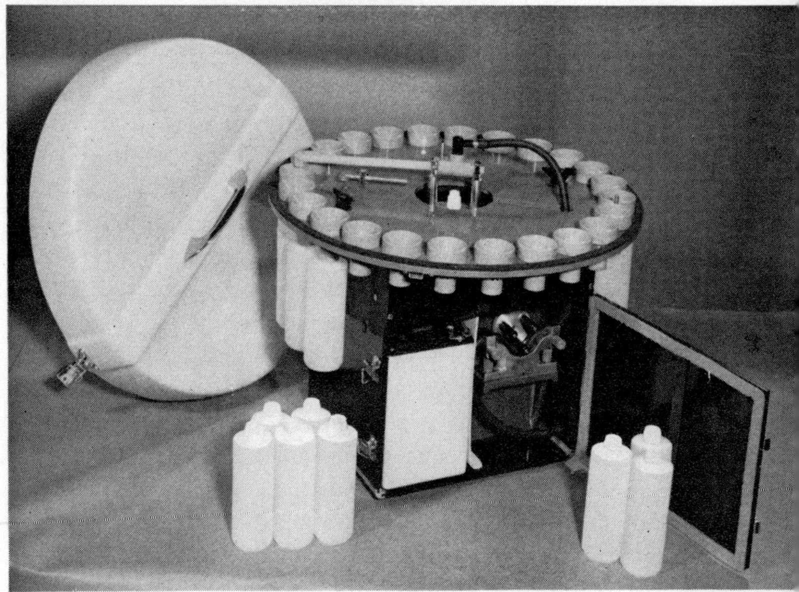

11. A battery-operated portable automatic sampler which provides 24 constant volume sample of liquid at regular time-controlled intervals

throughout the plant. This is the most reliable method of sampling for determining the nature and composition of crude sewage.

General schemes of treatment

The objects of sewage treatment are threefold: to convert the sewage into suitable end products, i.e. an effluent satisfactory for discharge to the local watercourse and a sludge in a disposable condition; to carry this out without nuisance or offence; and to do so efficiently and economically.

The fundamental process of purification is by settlement under gravity. The preliminary stages are physical, but, after settlement, the top liquor still contains non-settleable polluting matter. It is therefore subjected to a biological process which combines aerobic oxidation with conversion of the impurity to a settleable form which is removed by further gravitational settlement, sometimes followed by a stage of mechanical filtration or fine screening. The biological process carries out the oxidation of some of the organic matter present and may also oxidise part of the ammoniacal compounds.

The sludge may be disposed of in a variety of ways, depending upon local circumstances, but it is customary to dewater the sludge even if only to reduce the bulk to be transported. The removal of water from sludge becomes progressively more difficult. Most primary or mixed sludges can be dewatered down to a water content of about 92 to 94 per cent by gravity, but mechanical dewatering with or without the use of coagulant aids, or prolonged drying on open beds, is required if the water content of the sludge is to be reduced below 90 per cent.

Choice of method and design

Standardised package plants are sometimes employed for the treatment of small volumes of sewage from institutions or housing estates, but for schemes of larger size, the plant is designed specifically to suit the local conditions.

The site

The first consideration is the choice of site. While, for many years, engineers would go to considerable lengths to maintain a purely gravitational flow both to and through the plant, pumping equipment and control gear have now been developed to such a level of efficiency and reliability that designers have no hesitation in incorporating a pumping stage if this permits the choice of a better site.

In selecting a site, the designer has regard not only to the ease of sewering for a particular location, but also to the levels and configuration of the site; to safe and convenient access, both by private road and by the public highway; to the availability of land for future extensions; and, of course, to the suitability of the ground for excavation and construction work. Increasing attention is being paid to amenity, to the distance from houses, having regard to the direction of the prevailing winds, and to the appearance, visibility and general effect of the plant on the landscape.

The river authority may prefer a particular point of discharge for the effluent, which may not be to the nearest watercourse, or it may

12. An aerated grit removal tank. Air diffusers create a spiral flow, which induces the grit to collect in hoppers. The grit is then removed from the hoppers by air lifts

13. Simplex screen for crude sewage

offer a choice of effluent quality standards according to the location of the discharge. It has then to be decided whether to change the site or to convey the effluent across country.

A further question which may arise is whether or not to convey the sewage to an existing works instead of establishing a new plant. The additional sewering cost has to be equated with the savings effected by having a single treatment plant.

Preliminary treatment

Even where the sewerage system is nominally separate, it is prudent to make provision for the removal of grit from the incoming sewage. This can be effected by means of a small settlement tank from which the grit is removed by scraper gear and washed, or by passing the sewage along a constant velocity channel and removing the collected grit by a bucket elevator or a pump. More recently a grit chamber has been developed in which air is introduced along the bottom at one side, inducing a rotary motion which washes the grit *in situ* (as shown in photograph 12). This type of grit chamber has proved satisfactory in practice and now features in many new designs.

Comminutors are frequently used to eliminate the unpleasantness associated with screenings, but if the sewage is subsequently pumped difficulties due to the aggregation of rags, etc., sometimes arise. In such instances, and sometimes for other reasons such as the presence of excessive rags from textile factories, the employment of screens may be preferred. The mechanical clearing of screens presents some problems but these have been solved in a number of screens now available (photograph 13). Their operation is controlled automatically on a time or differential head basis. The screenings may be removed and buried or burnt, or they may be macerated and returned upstream of the screens. Where the flow to the works is by gravity, it is often advisable to incorporate a coarse screen with spacings of three or four inches (75 to 100 mm), in order to intercept occasional large debris such as sticks. If the screen is laid at a shallow angle to the horizontal, the flow will tend to push the debris out of the water and manual clearance will suffice. Whether the rags or the grit should be removed first is always open to question, but with the newer aerated grit chambers the decision is usually to install them before the screens.

Sedimentation

Manually desludged sedimentation tanks are no longer installed in Britain because the method is expensive in labour, unpleasant, gives rise to smell and involves the temporary loss of capacity. For these reasons also many existing horizontal flow tanks have been converted to mechanical desludging.

On some small works square vertical flow tanks with deep sludge hoppers are still installed but the excavation is often difficult and costly. The general practice is now to install relatively shallow radial flow tanks, equipped with mechanical desludging gear. In the earlier installations of rectangular tanks, a single transferable scraping machine served a range of tanks, but it is now the practice to provide each tank with its own machine (photographs 14 and 15). Considerable development has been carried out on mechanical desludging gear, and there is now available a variety of reliable machines covering all requirements, including those for quite small plants. In certain circumstances, the operation of the mechanism may be continuous, but, more often, it is intermittent and the frequency and duration of the desludging operation can be controlled automatically by devices for locating the level of the sludge or sensing the viscosity of the sludge withdrawn. Control is more reliable and objective when carried out by these devices than by workpeople.

14. Desludging equipment for rectangular tanks

15. Circular tank scraper

In terms of unit removal of pollution, sedimentation is, within its limits, cheaper than biological treatment, and it is uneconomic to allow the settled sewage to carry suspended solids forward unnecessarily and to deal with them at the biological stage. The sedimentation tanks need therefore to be operated at their maximum efficiency; an essential requirement is that the flow of sewage to each of the tanks must be properly apportioned over the full range of flow. This is by no means easy, and to achieve it calls for considerable skill and experience on the part of the designer. The design of the inlets and outlets of the tanks also has a material effect upon their efficiency and similarly requires detailed knowledge. In terms of value for money expended, the sedimentation stage justifies care and thoroughness, not only in the design, but also in the choice of the most suitable machinery and equipment to match the particular requirement.

Biological treatment

The choice lies between activated sludge and biological filters, both methods offering advantages and disadvantages. Apart from extended aeration plants (described in chapter 6), in general the smaller plants tend to use filters, while the larger the scale the more likely is activated sludge to be installed. Filters are usually higher in capital cost but lower in running costs than activated-sludge plants, and for most proposed schemes the estimated annual costs, i.e. loan charges plus operating and maintenance expenses, are not greatly different whichever method is considered. The decision is often made on grounds other than those of economics.

When it is necessary to produce an effluent of high quality, the area required for filters is about ten times that needed for an activated-sludge plant to treat the same BOD load. This ratio may be halved if recirculation or alternating double filtration techniques (described in chapter 5) are used. For plants required to treat large flows, the much smaller area required for activated sludge is the deciding factor and in recent years the maximum size of new plants employing biological filters has been decreasing. Few filter plants are now installed to cope with the flow from populations in excess of 50,000. Filters are more robust and recover more quickly from interference than do activated-sludge plants, and also require less skilled and close control. Filters are therefore more suitable for smaller works where technical facilities are limited.

The head required for an activated-sludge plant may be no more than a foot or so (about 30 cm), while the normal filter will require a fall of about eight or nine feet (2·5 m). Pumping settled sewage in a filter plant presents no difficulties, but it may affect the power and maintenance comparison.

There are other considerations. Filter installations are prone to breed flies, which may cause serious nuisance in the vicinity, and they are normally more conspicuous than activated-sludge plants. On the other hand, activated-sludge plants are often noisy; air compressors are difficult to silence completely, and mechanical aeration systems all give rise to noise. In residential areas this can be an important amenity consideration, especially at night.

The character of the effluent from the two processes varies. Filters loaded to produce a final settled effluent exerting a BOD of about 20 mg/l will usually oxidise up to about half the free ammonia present, whereas the effluent from an activated-sludge plant operated to the same standard will normally be devoid of oxidised nitrogen. The same activated-sludge effluent may, however, contain 15 mg/l or less of suspended matter, while the filter effluent is unlikely to carry less than 25 mg/l. The ability to vary the concentration of sludge carried allows a consistent level of performance to be maintained in cold weather in activated-sludge plants. Filters are more sensitive to temperature and nitrification and general oxidation levels fall during the winter. Unless filter distributing mechanisms are kept in operation continuously, they are liable to ice up during prolonged cold weather, while activated-sludge plants are much less subject to icing difficulties. Filters tend to achieve a more pronounced peak in efficiency in hot weather than activated-sludge plants. With the latter, care is necessary if denitrification in the final settling tanks is not to cause rising of the sludge.

Secondary sludges are more difficult to dewater than primary sludges, but surplus activated sludge is even more refractory than humus sludge. Furthermore, if surplus activated sludge, either alone or in admixture with primary sludge, is dried on open beds or lagoons, without prior digestion, it will give rise to serious odour nuisance. If it is intended to use sludge, either wet or dry, for agricultural purposes, then surplus activated sludge, which has a nitrogen content on a dry solids basis of about 8 per cent, appreciably higher than either primary or humus sludge, is the more valuable. The choice of biological process cannot therefore be made without full regard to the method of sludge disposal.

Two-stage biological treatment

There are occasions when two stages of biological treatment may be worth-while. When a sewage works requires extension, provided the hydraulic capacities of the existing units are adequate for the increased flow, an additional stage of treatment may be interposed. For example, for a short period, activated-sludge plant may be employed to reduce the strength of the settled sewage and so enable the filters to deal with the increased flow; alternatively, a high-rate filter can be used to give partial treatment before an existing activated-sludge or filter installation. Two-stage treatment is sometimes incorporated in the original design when difficult or toxic materials are likely to be present in the sewage and it is particularly important that nitrification and final effluent quality be maintained. It should, however, be remembered that two-stage treatment normally involves an extra stage of sedimentation, as described in chapter 5, and this affects costs. This can be avoided by the use of a high-rate filter, containing plastics media, to treat the macerated sewage before primary settlement.

Secondary sedimentation

It is now the general practice on all but the smallest works to employ mechanically desludged radial flow tanks for the treatment of both filter effluent and activated sludge.

Tertiary treatment

When a final effluent containing suspended solids of less than 30 mg/l is required some form of tertiary treatment is installed. On small works grass plots or lagoons are simple and reliable, but the area required makes them impracticable for larger plants. To deal with large flows, microstrainers are used or, for the more stringent standards, sand filters.

Sludge treatment

The dewatering and disposal of sewage sludge is still one of the most difficult problems in the whole field despite the range of methods which are available.

The greater part of the sewage sludge produced in Britain is subjected to anaerobic digestion. The primary advantages are the reduction of dry solid matter by about one-third and the avoidance of smell nuisance. The economics of the use of sludge gas from digestion plants for power production depends upon the availability and the price of an alternative supply of electricity.

The long-established practice of drying sludge on open beds has encountered labour difficulties which have been met by the development of mechanical aids extending to full automatic removal. Despite these improvements, the irregularity of the drying times has stimulated the search for methods which are independent of weather conditions and occupy less ground. There has been a revival of interest in the pressing of sludge conditioned either by chemicals or heat treatment, of vacuum filtration and of other mechanical methods.

The variety of methods now available to the designer is described in chapter 9, but the chief considerations in making a choice are:

(1) The reliability and consistency of the dewatering process.
(2) The lowest water content which can be achieved.
(3) The area of ground required.
(4) The extent and skill of the labour needed.
(5) The risk of smell nuisance.
(6) The economics.
(7) The method of ultimate disposal.
(8) The treatment of the separated water.

Summary

In designing either a new sewage treatment plant or extensions to one already in existence, the first essential is to know the flow, the nature and the composition of the sewage to be treated. Next, the local geography and circumstances must be ascertained in detail, together with the standards with which the effluent must comply. In the light of this information, the designer then makes his choice from the wide range of methods and equipment available, using his experience and skill to match them to the specific requirements.

D. H. A. PRICE

After graduating in biochemistry at Birmingham University in 1931, Mr Price went to the Birmingham Tame & Rea District Drainage Board as an assistant chemist. After several years he became assistant works superintendent at the Board's Minworth and Coleshill works. In 1939 he was appointed manager of the Rochdale sewage works, a post he held until 1950, when he went to the newly formed Severn River Board as chief chemist, subsequently becoming deputy pollution prevention officer. In 1955 he became a chemical inspector with the Ministry of Housing and Local Government, was promoted to senior chemical inspector in 1964, and took charge of the section on the retirement of Dr Key in 1967.

CHAPTER 5

Part 1: Biological Filtration

BY S. H. JENKINS, PhD, DSc, FRIC

Biological filtration has been used to purify sewage and industrial waste waters in Britain for at least 80 years. This process was developed to solve the difficulties associated with the treatment of sewage on land, one of the main limitations of which was the relatively large area of land required per unit volume of sewage applied. It was discovered that about 10 times the volume of sewage could be treated on a given area in unit time by passing it through a circular or rectangular bed of granular medium, supported on underdrains designed to allow access of air to the bed. It was found that the medium served as a support for bacteria and fungi whose activities brought about the purification of the liquid treated, provided that the rate of application was so chosen that the liquid passed over the film of micro-organisms in a thin layer. To achieve this in the most effective way, it was found necessary to provide a system of fixed or moving jets or sprays to spread the liquid uniformly over the upper surface of the bed.

The process spread steadily to all parts of the world and is now extensively used in the treatment of a very wide and increasing range of types of waste water. With appropriate attention to mechanical equipment, a well-designed plant will perform satisfactorily for a very long time—indeed, virtually for the lifetime of the medium. When one British plant that had been in continuous use for 70 years was demolished in 1968 to make room for improvements to a works, the medium looked good enough to last another 50 years. Practical developments in biological filtration have been very largely the outcome of empirical experimentation, and it is these practical developments which have tended to run ahead of fundamental theoretical explanations of the causes of their effectiveness that will be emphasised in this chapter.

In Britain, one of the most common requirements for filters has been the production of effluents of high quality, satisfying the Royal

Commission 30/20 standard from sewage or industrial effluents of similar strength. The simplest conventional form of filter is usually about six feet (1·8 m) deep. Almost any type of mechanically strong and durable granular material can be used for the medium, broken rock, gravel, clinker and slag being among those commonly employed, but it must be uniformly graded and of such a shape as to give a large proportion of voids when assembled. The grading usually chosen for conventional treatment is within a size range classified as $1\frac{1}{2}$–2 inches in accordance with specifications prescribed by the British Standards Institution. Material of this grading has a specific surface area of 25 to 35 ft^2/ft^3 of volume (80 to 110 m^2/m^3), and the proportion of voids in the assembled bed is normally in the range 45 to 55 per cent of the total volume. Liquid is usually applied to circular filters by means of a rotating system of radial sparge pipes; in cases of rectangular beds, the sparge pipes are arranged across the filter and travel backwards and forwards along its length. In some designs, movement of the distributors is actuated by reaction to the flow of liquid, but electrically-driven distributors are also used (photograph 16), particularly in some of the larger works. Before final discharge, effluent from the filter is usually passed through a tank where suspended matter (humus) flushed from the bed is removed by settlement.

16. Electrically-driven distributors on bacteria beds

The purifying action of a filter depends essentially on the utilisation of biologically degradable constituents of the waste water by bacteria and fungi within the film, as nutrients for their growth. The term 'filter' is essentially a misnomer, though undoubtedly physical processes such as adsorption, coagulation and entrainment play a part in the removal of suspended matter. The detailed composition of the biological film depends a great deal on the operating condition and nature of the waste water treated, but it is usually populated by algae, in those areas exposed to light, by protozoa and by macro-fauna, including especially the larvae of flies and worms. In a filter which is working satisfactorily, the film, though continually growing, does not accumulate to a point where it would block the interstices, but is continually breaking away and being released to the effluent as humus. A vital role in keeping the interstices of the filter open is played by macro-fauna, which devour the biological film and release residues of it as faecal pellets which form a large part of the humus discharged. Their activity increases with increasing temperature, and it is partly for this reason that the amount of humus discharged from filters and their overall performance varies seasonally.

The rate of removal of organic matter from the liquid as it passes through a filter tends to be roughly proportional to its concentration and thus the reduction in this concentration occurs most rapidly in the upper regions of the filter. Nitrifying bacteria will commonly develop in conventional filters treating sewage and some other waste waters containing ammonia, but, partly because higher concentrations of dissolved oxygen are required by these organisms than by the heterotrophic bacteria which oxidise carbon, rates of nitrification tend to be higher in the middle and lower regions of a filter than at the top, where the demand for oxygen is highest. Nevertheless, it is usually possible to achieve virtually complete nitrification of ammonia.

One of the chief difficulties in operating filters is that excessive quantities of biological film can accumulate, thus blocking the bed and causing ponding at the surface; this leads to channelling and to a deterioration in effluent quality. The quantities of film in a filter can be measured *in situ*, using a modified form of soil-moisture meter (photograph 17) developed at the Water Pollution Research Laboratory, and studies indicate that, with conventional single-pass filters, a critical condition appears to be reached when about 50 per cent of the volume of voids is occupied by film. Other things being

17. Measuring the moisture content of a percolating filter by neutron scattering

equal, the liability of filters to pond increases with increasing organic loading and this imposes a practical limit on the maximum loading. With the conventional type of filter under consideration, the maximum daily loading for consistent production of a 30/20 effluent is about 0·2 lb BOD/yd^3 (0·1 kg/m^3), although instantaneous loadings up to about three times this figure can be tolerated. Inevitably, much attention has been given to methods of circumventing this limitation and British engineers have played a leading part in the development of a number of procedures which allow this to be done.

Recirculation

Recycling of effluent so as to dilute the incoming waste water serves to reduce the rate of growth of film, to produce a better flushing action for removal of loose film, to promote more effective contact with the available surface of the medium and a more uniform distribution of film with depth. Nitrification is often reduced when recirculation is employed, probably as a result of increased competition from heterotrophic bacteria in the lower parts of the filter, but two or three times the normal loading for single-pass filters can be treated with the production of a satisfactory effluent.

Alternating double filtration

In this process, which is widely practised in Britain, especially ir large works, two filters are operated in series. The waste is applied a a relatively high rate to the primary filter, and its effluent, afte settlement, is then passed to the secondary. At intervals, ranging from daily to weekly, the order of the filters is reversed. When a filte is serving as a primary, film grows rapidly and would eventuall choke the filter, but when it is switched to the secondary role, the film is fairly rapidly depleted. Larger distributors, more complicatec pipework and additional sedimentation capacity are required a compared with the single-pass filter, but three to four times the volume of liquid can be treated on a given volume of medium, anc overall costs are generally lower than for single filtration.

Control of distribution

The way in which liquid is applied can affect the tendency of filter to pond. Distribution in the form of a fine spray commonly cause heavy surface growths, and it is usually best to distribute as a numbe of strong jets even if these do not entirely wet the surface; there is alsc an optimum dosing interval. On circular filters, distributors of th type driven by the reaction of jets often revolve too quickly and a some works brakes are fitted. Electrically-driven distributors facilitat control of dosing at the optimum interval and are being used to ar increasing extent.

S. H. JENKINS

Dr Jenkins graduated in chemical technology at Manchester University an later took a Master's degree in fuel technology. His experience with effluen problems began when he entered the textile industry. This took him to Rotham sted Experimental Station where, for 11 years, he worked on investigations int methods for purifying beet-sugar wastes, the effluent from the milk industry, an anaerobic studies on fat degradation. In 1938 Dr Jenkins joined the Birming ham Tame & Rea District Drainage Board, and in 1966, when it became par of the Upper Tame Main Drainage Authority, he became its chemist. He wa responsible for creating a department to control trade waste discharges to th sewer, and the construction of laboratories that rank among the best in th world in the field of water pollution control. Although now retired, he remain as consultant to the Upper Tame Authority and in waste water treatment.

CHAPTER 5

Part 2: Plastics Media in the Treatment of Domestic Sewage and Industrial Wastes

BY P. N. J. CHIPPERFIELD, BSc, PhD

The limitations to the applications of biological filtration imposed by the tendency of conventional media to clog at high organic loadings, and their inherently large bulk and weight in relation to specific surface and voidage, have stimulated a search for improved forms of medium. A major breakthrough came with the introduction of packings fabricated from synthetic plastics materials. These generally comprise an ordered arrangement of vertically orientated spaced sheets, of which the individual proprietary design has different geometrical complexity; two examples of designs developed in Britain are shown in photographs 18 and 19.

The basic conception of plastics media arose independently, and more or less simultaneously, on both sides of the Atlantic, largely in response to the need for a more economic and technically satisfactory solution to the problem of treating strong industrial wastes than could be provided by the use of the activated-sludge process or conventional trickling filters, and also as a consequence of the rapid advances in plastics technology made during and immediately after the 1939–45 war. Although the commercial introduction of plastics packings occurred somewhat earlier in the United States, it would seem that applications have been more numerous and varied in Britain, where, for example, over 60 plants employing one type of British packing are now operating or being built and another type of packing is undergoing pilot trials in several places. Judging from the literature, considerably more attention has been given in Britain than in the USA to the use of plastics media for multi-stage operation in the treatment of industrial wastes (particularly strong ones), either before their discharge to municipal sewers, or as a precursor to further purification to a 30/20 standard or better, and for the 'roughing' treatment of domestic sewage in order to remove overloads from existing sewage works, or to reduce the costs of new sewage works.

18/19. Two examples of plastic filter media: Flocor (above) and Crinkle-close Surfpac (below

Basic characteristics of plastics media

Many of the constructional problems and limitations associated with the large bulk weight of conventional media, and the operational problems, such as clogging, which can occur with their use for high-rate treatment, have been overcome in the design of plastics packings. In comparison with conventional media, these can be designed to be:

(1) capable of removing a greater weight of BOD/unit packed volume;
(2) able to operate at higher hydraulic loadings/unit volume and unit area;
(3) of such configuration as to induce the liquid to flow uniformly over the available surface of the 'bios' (microbial film) in as thin a turbulent film as possible, in order to promote the maximum rate of transfer of substrate and oxygen to the bios;
(4) sufficiently open in structure to avoid blockage by the accretion of solids and to ensure an adequate supply of oxygen without recourse to forced aeration;
(5) sufficiently light in weight (even when loaded with bios) to enable a significant reduction to be made in the civil engineering costs of plant construction;
(6) no more expensive in terms of cost/effectiveness;
(7) sufficiently strong structurally to bear their own weight and the weight of overlying layers of medium, together with the attached growths of bios and included water;
(8) biologically inert, neither attacked by, nor inhibiting the growth of, the bios.

Materials suitable for use as plastics packings

In physical terms, an ideal plastics medium must be of low bulk density, capable of ready fabrication into desired shapes in order to obtain the greatest possible area of surface on which bios can grow per unit weight of material, and must possess surface characteristics that promote the adhesion of bios. Existing materials that fulfil at least the first two criteria are almost entirely synthetic polymeric substances, the surface area/unit weight of which can be expanded by mechanical or other techniques.

Many plastics are unsuitable because of cost or chemical and physical properties. Polyvinylchloride (PVC), whose durability in many kinds of environment has now been demonstrated over more than 30 years of general use at normal temperatures, is particularly satisfactory. Nevertheless, even with PVC, care has to be exercised in designing a packing to ensure that adequate strength is obtained by the use of sheets of the correct thickness and in the correct assembly, to ensure support of the weight of packing together with the bios. Polystyrene and polystyrene blends are used for some types of media, but such materials have certain disadvantages in regard to brittle strength, chemical resistance, and flammability.

Honeycomb-type packings made of plastics-coated or partially impregnated paper, fabricated originally for water cooling purposes, are not suitable for biological filters, since they appear to fail mechanically after comparatively brief periods of service (1 to 12 months); the impregnation or coating processes thus far employed fail to provide complete protection against the absorption of water by the paper, causing softening and disintegration or collapse.

The practical design of plastics packings

The function of a packing is to provide a mechanical support for the bios in such a way that this is readily accessible to the contaminants in the waste water to be treated and to the penetration of adequate atmospheric oxygen. One might expect that the percentage removal of BOD from a given liquid would increase with increase in the amount of bios with which it had been brought into contact. Furthermore, it would seem logical to postulate that the amount of bios per unit volume of packing should be, under stable equilibrium conditions, directly proportional to the 'specific surface' (i.e. the surface area/unit volume) of the packing, and thus at a given hydraulic load BOD removal should be proportional to the specific surface.

Considerable investigation of packings of different geometric form and specific surfaces has shown that, in practice, this is not always so, largely because in some cases all the included surface is not available for the growth of bios owing to incomplete 'wetting' of the surface, leading to by-passing of a significant proportion of the surface by 'streaming' or 'channelling'.

To some extent, this failure to utilise the entire included surface can be overcome by increasing the hydraulic load, since for each particular packing a 'minimum irrigation rate' (analogous to the concept of the 'minimum wetting rate'—a function of the peripheral length per unit area of cross-section of the packing—employed in chemical engineering in the design of water cooling or gas adsorption packings) can be postulated. In practice, however, the hydraulic régime in the packing is of primary importance, since ill-designed packings of high specific surface can still exhibit poor utilisation of included surface even at high hydraulic loads.

Experience has shown that packings consisting of straight (and especially narrow) vertical tubes or channels are particularly subject to hydraulic streaming or channelling, which leads to a localised growth of bios and eventually to complete ponding of the packing. This problem is significantly aggravated by poor (non-uniform) distribution of liquid at the top of the filter, particularly where little opportunity is afforded for redistribution from layer to layer within the filter. Satisfactory distribution of liquid at the top of a filter poses considerable problems when media of high specific surface are employed.

In general, there appears to be a limiting velocity of the falling film of liquid above which the bios film tends to be of constant thickness, and below which local heavy accumulations of bios can occur, which can themselves increase the tendency of the packing to block since, when they exfoliate, the clearances within a narrowly spaced packing may be too small to allow them to be voided.

Biological treatment on plastics packings

All well-designed plastics packings show the same general characteristics when employed in high-rate biological filters, although very considerable differences in actual BOD removal efficiencies at different loads, in limiting hydraulic loads, and in ponding tendencies, etc., are demonstrated in practice between different designs and types. The general characteristics may be summarised as follows:

(1) Percentage removal of BOD tends to decrease with the increasing BOD loading, though the changes produced by a given increase diminish as loading is increased. The removal obtained at a given loading varies with the intrinsic treatability of the waste water, but is often at least 90 per cent at loadings up to 2 lb/yd^3 d (1·2 kg/m^3 d), decreasing to around 50–60 per cent at from 7 to 20 lb/yd^3 d (4·1–11·7 kg/m^3 d), depending on the nature of the waste water.

(2) Multi-stage treatment has proved practicable and effective for many kinds of industrial wastes (see Table 1) and for domestic sewage.

(3) Increasing the hydraulic loading above an optimum specific for a given packing produces a decrease in the percentage

BOD removal, the relationship being linear over a wide range of rates. At the same time, the weight of BOD removed per unit packed volume of filter increases. A limitation is the 'flooding' of the packing, the irrigation rate at which this occurs depending on hydraulic design.

(4) Further, at a given hydraulic loading, within a range specific to the packing, the percentage removal increases as the packed depth of the filter is increased. However, the use of deep filters is only technically justified where space is limited, where recycle is necessary, or where a relatively high degree of purification is required in once-through operation. Where the maximum weight of BOD removed per unit packed volume of filter is required in once-through operation without a high degree of purification (e.g. in the 'roughing' treatment of domestic sewage and similar wastes with BODs of moderate strength), shorter filters show a marked advantage. In practice, filters deeper than 24 feet (7·3 m) are rarely required technically and even more rarely are economic. The contention that the use of tall filters enables considerable savings to be made in the total volume of packing employed to fulfil a given duty would appear on critical analysis to be erroneous.

(5) Recycle, except where necessary to maintain the optimum irrigation rate, or to dilute wastes which otherwise could not be treated, appears, if anything, to decrease the efficiency of removal (unlike the effect in conventional filters).

(6) In properly operated plastics packed roughing filters, the bios is relatively uniformly distributed over the entire depth of the filter. In a well-designed packing, even at high BOD loads, occlusion of the voids by bios growths is rarely greater than 10 per cent.

(7) A roughing unit exhibits a high degree of resistance to shock loads, sudden increase in BOD load or hydraulic load producing no significant reduction in percentage purification.

(8) The quantity of settleable solids produced by roughing filters increases with the applied BOD load although the amount produced depends also on the nature of the waste. In practice it has been found that 15 to 20 per cent of the BOD removed is converted to humus solids.

(9) Where recycling is used, it is frequently advantageous to remove at least a proportion of the humus solids from the recycle liquor by means of 'in-loop' or interstage settlement, or other methods.

Practical applications of plastics media in Britain

Plastics packings are employed in two main ways: (a) for high-rate treatment of strong industrial wastes where recycling is employed to maintain an adequate flow of liquid; (b) for high-rate treatment of wastes of lower BOD (especially domestic sewage) at higher hydraulic loadings, without recycling.

Many of the plants operated in Britain fall into class (a). Data for some examples of the use of one type of packing are given in Table 1. In all these instances, a conventional filter plant to carry out the same duty would have been much more costly, because of the low loadings that would have had to be employed, while in many instances treatment by activated sludge was technically impracticable, for reasons such as low sludge index, loss of sludge or sludge bulking.

Considerable evidence is accumulating that, besides being robust, plastics filters are often less susceptible to shock loadings than activated-sludge plants; they are also largely self-regulating and able to work for long periods with little attention (other than for desludging settlement tanks). In the majority of applications, they can be erected at a comparatively low capital cost and operated at an economical running cost, both in manpower and in electric power requirements. For the same degree of purification, the pumping costs for a plastics-packed filter appear to be less than one-tenth of activated-sludge power costs. Two typical British plastics-medium-packed plants are shown in photographs 20 and 21.

Plastics 'roughing' packings can be used in three major ways in the treatment of sewage:

(1) to relieve the overload on an existing sewage works by removing an appropriate proportion of BOD, so that the partially treated sewage can then be purified to the required standard in the existing biological plant;

(2) to form a first high-rate stage in a two-stage plant, employing conventional granular medium or activated sludge in the second stage;

(3) to provide complete treatment to the required standard in a single- or two-stage plant.

Uses (1) and (2) seem likely to provide many opportunities in Britain for economies to be made in the capital cost of biological treatment plant. Use (3) is only likely to prove economic for sewage treatment in Britain where BOD removals of less than 85 per cent are required. In some countries where the sewages are weaker than those in Britain such treatment has been shown to be sufficient to produce an effluent having a BOD of less than 20 mg/l.

TABLE 1

Performance of some plastics-medium-packed plants in Britain treating strong industrial wastes (200–5000 mg/l BOD)*

Type of waste	Average loading		% removal
	$lb/10^3ft^3$ d	kg/m^3 d	
Whisky distillery (MS)	109[1]	1·75	93[2]
Whisky distillery (MS)	83[1]	1·33	87[2]
Synthetic resin (MS)	34[1]	0·55	94[2]
Synthetic fibre (MS)	37[1]	0·59	97[2]
Phenolic waste (SS)	115	1·84	94
Coke-oven (MS)	111[1]	1·78	91[2]
Synthetic fibre (SS)	83	1·33	76
Fruit & vegetable canning (SS)	102	1·63	80
Brewery (MS)	170[1]	2·72	86[2]
Cheese factory (SS)	315	5·05	89
Coke-oven (SS)	175	2·89	79·5
Textile dyeing (SS)	200	3·20	63
Pharmaceuticals (MS)	322[1]	5·16	86[2]
Meat packing (SS)	230	3·68	76
Coal processing (SS)	445	7·13	50
Yeast waste (SS)	370	5·93	67
Cider (SS)	78	1·25	67
Porteous Process (sewage sludge heat treatment process) (SS)	320	5·13	60

* 'Dissolved' BOD, i.e. BOD after settling in laboratory for one hour.
MS = cascade multi-stage. SS = single-stage.
[1] = average overall loading for multi-stage plants.
[2] = overall removal for plastics-packed stages only.
Recycle used in all cases to give irrigation rate of 0·5 gal/ft^2 min (24·4 l/m^2 min).

P. N. J. CHIPPERFIELD

Dr Chipperfield is scientific adviser on water and effluents to Imperial Chemical Industries Ltd, and has been manager of their Brixham Research Laboratory since 1958. The Laboratory provides an advisory and investigatory service to ICI and to British industry in general, on a commercial basis. During the past few years, the Laboratory has been particularly concerned with biological treatment of industrial waste and sewage on plastics media, together with ancillary work on heat and mass transfer.

0. Plastics-packed treatment plant for a whisky distillery effluent

1. Plastics-packed treatment plant for a synthetic fibre production effluent. Note the corrugated VC cladding to the towers

CHAPTER 6

The Activated-Sludge Process

BY G. AINSWORTH, BSc, FInstWPC, FIPHE

Where large volumes of waste water are to be treated to a high standard, or suitable land is scarce or expensive, treatment by conventional biological filtration, as described in chapter 5, may occupy too much land. Considerable economies in land requirements and initial installation costs, if perhaps at the expense of higher operating costs, can often be achieved by adopting the activated-sludge process.

This title is usually applied to any treatment system in which waterborne wastes of domestic or industrial origin are aerated in the presence of flocculent cultures of micro-organisms, the so-called 'activated sludge', freely suspended in the liquid. The avoidance of any inert supporting structure for the organisms enables very compact units to be used.

As mentioned in chapter 1, the process originated at Manchester, England, in 1913, being discovered by Fowler, Ardern and Lockett, members of the staff of that city's Davyhulme Sewage Works. It is recorded that these scientists decided not to patent the process but, in Dr Fowler's words, 'to give it to the world' so as to facilitate its large-scale application. That generous decision resulted in the establishment of highly efficient intensive treatment techniques which may be as readily applied to the sewage from small communities as to that from the largest cities of the world.

Most waste waters amenable to treatment by the process contain an inoculum of micro-organisms and, on aeration, those adapted to the prevailing environmental conditions will grow. A proportion of these organisms will usually be found to clump together, and the flocculated mass can be separated by sedimentation. Further quantities of the waste water can then be added, and the procedure repeated until a sufficient concentration of flocculent activated sludge has been built up to permit operation under continuous flow conditions. The concentration usually required in the mixture of sludge and waste water is within the range of 1500 to 8000 mg/l; the time needed to acquire it may be as little as a few days.

In the simplest form of conventional continuous flow process, the settled waste water is mixed with the activated sludge and aerated, usually for a period of between one and thirty hours, depending on the treatment required. Purification takes place by a series of processes of which the most important is the utilisation of components of the waste water by bacteria, which convert a proportion into new bacterial cells and oxidise the remainder to provide energy for cell synthesis. After aeration, the sludge is removed from the purified liquid by sedimentation, and most of this sludge is recycled to the inlet end of the aeration unit.

Since the process of purification results on average in a net increase in the concentration of sludge during its passage through the aeration unit, excess sludge must be bled off continuously or intermittently, so as to maintain the concentration in the mixed liquor at a satisfactory level for efficient operation. If one considers the mixture of waste water and sludge to be passing through the aeration units rather like a piston, changes in the composition of both liquor and sludge take place as the mixture progresses from inlet to outlet. These changes can be thought of as occurring in distinct phases, in which different processes predominate, though in fact there is no sharp distinction between them, especially since there is always some degree of longitudinal mixing of the tank contents. Most waste waters amenable to treatment by the process contain materials of which a proportion becomes adsorbed or absorbed by activated sludge almost immediately after mixing at the inlet end of the aeration units. Utilisation of the biodegradable components, both of this material and of that remaining in the liquid phase, by micro-organisms in the sludge, also begins almost instantaneously. The rate of growth of micro-organisms on such substances tends to rise with increase in their concentration up to a certain level, beyond which the growth rate remains roughly constant, probably because the enzyme systems governing rate-limiting processes in cell metabolism are effectively 'saturated'.

In the initial stages of aeration, the concentrations of nutrients are at their highest, and thus the rates of microbial growth, the rates of destruction of oxidisable matter, and the rates of consumption of dissolved oxygen by the respiring organisms are correspondingly high. As the substrates become used up, the rate of multiplication of the micro-organisms decreases, and both the rate of oxidisable matter and the respiratory oxygen demand begin to fall.

Upon exhaustion of the substrate, the organisms respire endogenously, a process resulting in a gradual auto-oxidation (or aerobic digestion) of cellular material, followed by death and disintegration. During this phase, the rate of consumption of dissolved oxygen declines still further. Other things being equal, the longer the duration of aeration, the greater will be the degree of aerobic digestion of sludge, and thus the smaller will be the net production of surplus sludge.

Various other factors also affect sludge production. For instance, it tends to increase with increasing organic loading and to decrease with increasing temperature and with increasing concentration of activated sludge maintained under aeration. Operating variables, particularly the nature of the waste water added, also determine the microbial population which will develop in the sludge, and hence its purifying ability and physical properties. Part of the influence of operating variables on the microbial population results indirectly from their effect on sludge production. This is because a fraction of the micro-organisms in the system is continually removed as a result of the withdrawal of surplus sludge. Without going into details, it can be shown that, unless the percentage rate of growth of individual species of micro-organisms exceeds the percentage rate of removal of activated sludge, an effective population of the organisms in question cannot be maintained in the system.

This situation has its most noticeable impact on the ability of sludges to oxidise ammonia, since the nitrifying bacteria which bring this about grow rather slowly by comparison with the organisms which metabolise carbonaceous matter, and nitrifiers cannot be retained in the system under conditions giving high percentage rates of sludge production. The Water Pollution Research Laboratory at Stevenage has been able to demonstrate both mathematically and practically the limiting relationships between activated-sludge concentration, aeration period, temperature, dissolved oxygen, and pH value which govern the possibility of oxidising ammonia in sewage and other waste waters.

Plants installed in Britain for treating sewage or waste waters of similar character to the high standards required for discharging to rivers are usually designed so as to provide retention periods of from four to ten hours in the aeration units. For some applications, such as for instance, discharge of effluents near the mouth of estuaries, or when the plant forms the first stage of a two-stage process, only partial purification is needed. This can be achieved by employing high-rate plants operating with high organic loadings per unit of plant capacity; and, for a given type of waste water, much shorter retention periods than is necessary to produce a fully nitrified effluent.

High-rate treatment usually requires high concentrations of activated sludge (4000 to 8000 mg/l) in the aeration tank to perform the necessary oxidation in the short detention period provided, and this generates a high oxygen demand, which must be met by providing aeration of adequate intensity. The savings in capital cost resulting from the high rate of removal of polluting matter per unit volume of plant tend to be offset to some extent by increased production of surplus sludge, and by the fact that this sludge is sometimes more difficult to dewater than sludge from conventional plants.

A form of treatment known as the contact-stabilisation process (photograph 22) takes advantage of the fact that, in many waste waters, much of the impurity is present as suspended matter of a form readily adsorbed or entrained by activated sludge, and many of the soluble constituents are readily oxidisable. The waste water is aerated with sludge in a contact zone for a short period (one to one and a half hours), by which time most of the unoxidised impurity

22. Contact-stabilisation unit treating domestic sewage from a small community

(other than constituents such as ammonia) is concentrated in the sludge. Sludge is then separated by sedimentation and aerated in a separate compartment, usually for several hours, before being finally returned to the contact zone. The concentration of sludge in the aeration compartment is much higher than in the mixed liquor; thus, the size of compartment needed to provide aeration of sludge for a given period is much smaller than that which would have been required had the sludge not been separated. The process is suitable for use at large works, though in Britain its application has been largely confined to treatment of relatively small volumes of water-borne waste from small communities or factories. Prefabricated plants are marketed for this application, for which factory-built extended aeration plants are also widely used. Small plants working on a similar principle have also been developed by British firms for use in ships (photograph 23).

Usually, in the latter type of plant, the waste water is not settled before admission to the aeration compartment, although coarser solids may be comminuted or removed on screens. The prolonged aeration period provided, usually longer than 24 hours, reduces the final quantities of surplus activated sludge requiring disposal

23. Prefabricated activated-sludge unit treating waste waters on board ship

Whatever system of operation is decided upon, the oxygen requirements of the process will be supplied by direct diffusion of air bubbles, by air entrainment using mechanical agitators, or by combinations of the two. Many different types of aeration systems are available from British manufacturers; the following summary can give only an outline of some of the equipment available.

Air diffusion systems

The very first activated-sludge units employed diffused-air aeration, initially through open pipes and subsequently through porous diffusers. The modern counterparts of this method bear only superficial resemblance to the earlier treatment systems. Continuous development work has produced very satisfactory diffusion systems, matched by efficient air filtration and compression equipment.

The most common British air-diffusion system is one developed by Activated Sludge Ltd, in which aeration is effected by finely dispersed air bubbles. The original porous-plate diffusers set in

24. Base of fine-bubble air diffusion unit, employing 7-inch diameter porous diffusers

cast-iron trays have been replaced by easily removable 7-inch (0·18 m) diameter porous domes mounted in a high-density arrangement along the top of a series of air mains fixed to the bases of flat-bottomed tanks, which are cheap and easy to construct (photograph 24). With efficient air filtration and supply, such units can operate for 10 years before the diffusers need cleaning by chemical treatment or by firing in a kiln. As an example of a typical modern plant, the Maple Lodge Works of the West Hertfordshire Main Drainage Authority (photograph 25) utilises fine-bubble aeration of this type to produce an effluent which generally contains less than 15 mg/l suspended solids and has a BOD below 15 mg/l, with averages much lower than these values.

The aeration units treat 30 lb BOD/1000 ft^3 of tank capacity (0·48 kg/m^3), 900 ft^3 of air/lb BOD (56 m^3/kg) being supplied. In addition to removing about 2 lb BOD/kWh (0·9 kg/kWh), virtually the whole of the nitrogenous constituents of the 26 million gallons a day (118,000 m^3 d) of sewage are oxidised, there being only traces of ammoniacal nitrogen in the final effluent, which subsequently contributes to London's water supply.

Although they have somewhat lower oxygen-transfer efficiencies coarse-bubble aeration systems, employing perforated tubes or open-ended pipes, can have certain advantages. Larger volumes of air are required, but air filtration is not necessary, and simpler, more efficient compressors can generally be used, because of the lower air pressures required. Coarse-bubble aeration is particularly applicable in small extended-aeration systems, where skilled maintenance may not be readily available. However, the Jet Aeration Process manufactured by William E. Farrer Ltd is also successfully applied on large-scale installations. It uses horizontal tubular grids to distribute large volumes of coarse air bubbles from the bases of flat bottomed tanks. At the city of Coventry's Finham works, this type of aeration has been used to give partial treatment to sewage, which is then further purified by biological filtration. Investigations have shown that the aeration plant can produce effluents having less than 20 mg/l each of BOD and suspended solids, within detention periods of less than four hours.

The Freehold Works of the Upper Stour Main Drainage Authority (photograph 26) also use this process to treat sewage from 75,000 people, containing a high proportion of metal-finishing wastes. A nominal detention period of six hours is provided, and though rapid-gravity sand filters have been constructed to provide tertiary

25. One of the 25-feet (7·6 m) wide aeration units at the Maple Lodge Works of the West Hertfordshire Main Drainage Authority which utilises fine-bubble air-diffusion

26. Coarse-bubble aeration units at the Freehold Works of the Upper Stour Main Drainage Authority, employing Jet aerators

27. The Simplex High Intensity Aeration Cone, which consists of a rotating funnel-shaped agitator, equipped with vanes on its upper surface surmounting an uptake tube situated in the centre of each aeration pocket

28. Simplex High Intensity Cones treating industrial sewage at the Aldwarke Works of the County Borough of Rotherham

effluent treatment and prevent the effluent analysis exceeding 20 mg/l suspended solids and 15 mg/l BOD, their use has so far proved unnecessary.

Apart from those in various proprietary extended-aeration units, other coarse-bubble systems are available from a number of British manufacturers.

Mechanical aeration systems

The compressed air used in the foregoing systems serves two functions: it provides oxygen, thus enabling purification to proceed, and it keeps the activated sludge circulating through the liquid being treated. In the mechanical aeration systems, atmospheric oxygen is introduced into the liquid and the activated sludge kept in suspension by an agitator rotating at or near the surface of the aeration tank.

The Simplex High Intensity Aeration Cone, developed by Ames Crosta Mills & Co Ltd, consists of a rotating funnel-shaped agitator, equipped with vanes on its upper surface, surmounting an uptake tube situated in the centre of each aeration pocket (photograph 27). Circulation and aeration are produced by the aeration cone throwing liquid from the draught tube over the tank surface, entraining and dissolving oxygen in the process.

The intensity of aeration can be controlled by varying either or both the rotation speed of the cone and its degree of immersion in the liquid. When the extensions to the city of Manchester's Davyhulme Works are completed, a dry weather flow of 82 million gallons (370,000 m^3) a day of one of the most intractable industrial sewages in Britain will be treated by 88 six-feet (1·83 m) diameter Simplex Cones and 49 ten-feet (3·05 m) diameter cones. Rotherham's Aldwarke Works (photograph 28) treats a dry weather flow of 3·8 million gallons (17,300 m^3) a day of industrial sewage by 32 six-feet (1·83 m) Simplex Cones to produce a non-nitrified effluent having less than 30 mg/l suspended solids and 20 mg/l BOD. The loading of the aeration plant is 30 lb BOD/1000 ft^3 of tank capacity (0·48 kg/m^3) and 2·8 lb BOD/kWh (1·28 kg/kWh) are removed.

Simon-Carves Ltd have developed the Simcar aerator for chemical process engineering, which is now available for effluent treatment. This is a cone with eight blades on the underside (photograph 29); rotation produces intense aeration beneath the cone and aerated liquor is thrown in a low trajectory towards the sides of the tank; it is then replaced by liquor rising from the bottom of the tank, thereby inducing a powerful circulation. No draught tube is used.

29. A Simcar aerator in empty tank showing aeration blades on underside of cone

The London, Ontario, plant in Canada utilises 72 Simcar aerators, each 6 ft 8 in (2 m) in diameter, to treat a dry weather flow of 12 million gallons (54,500 m^3) a day. At the Newbridge sewage works of Midlothian County Council, Scotland, 20 single-speed aerators of 4 ft 8 in (1·4 m) diameter produce an effluent having much less than 30 mg/l suspended solids and 20 mg/l BOD from a dry weather flow of 2,750,000 gallons (12,500 m^3) a day, with maximum design flow reaching 9,900,000 gallons (45,000 m^3) a day. The loading on the plant is equivalent to 45 lb BOD/1000 ft^3 (0·72 kg/m^3) and 2·8 lb BOD/kWh (1·28 kg/kWh) are removed (photograph 30). These particular units are fitted with parabolic weirs which increase the aerator immersion and intensity of aeration in proportion to the flow.

Both Simplex and Simcar systems may be applied as single-aerator units, or in rectangular tanks using series of aerators. Each aerator is available in several sizes for incorporation in different sizes of tank. Further mechanical aerators are built under licence in Britain by a number of manufacturing firms.

Although produced primarily for use in activated-sludge units, many of the aeration devices mentioned above have also proved of great value in maintaining satisfactory dissolved-oxygen levels in maturation ponds, lakes and other polluted stretches of water. In such applications, they are usually supported by a floating platform.

Design considerations

When deciding which aeration system and mode of operation to adopt, the designer of an activated-sludge plant will need to evaluate several factors in addition to the characteristics of the influent and quality of effluent desired. The final choice will generally be a compromise, but the following specification lists some of the desirable features to be aimed at and the factors to be taken into account:

Design target	*Contributory factor to be considered*
Minimum installation cost:	Land requirements: area and depth.
	Size and simplicity of civil, mechanical and electrical engineering required.
	Number of aeration devices and their means of operation.
Minimum operating cost:	Aerator efficiency (including driving arrangements where applicable).
	Effects of detergents on oxygen transfer.
	Maintenance requirements: reliability, durability (materials of construction), accessibility.
	Effect of climatic conditions.
Minimum side effects:	Impact on other treatment processes (e.g. preliminary treatment required), secondary sludge production.
	No floc disintegration: peripheral velocity of mechanical aerator $<$ 30 ft/sec (9 m/sec).
	Adequate circulation velocity to prevent deposition: preferably $>$ 1 ft/sec (0·3 m/sec).
	Detergent foam production and ease of suppression.
	Spray drift.
	Noise.
Adaptability:	Ease of increasing treatment capability.
	Effects of fluctuations in volume and/or load.
	Ease of automation or modifying pattern of operation.
	Effects of power failure on subsequent aeration efficiency.
	Effects of subsidence of site on air distribution or aerator operation.

Many of the points listed will be self-evident and no elaboration should be necessary. Some, however, deserve further comment.

The presence in sewage of surface-active agents, such as domestic detergents, may not only generate copious quantities of embarrassing foam on aeration tanks but can also interfere with oxygen transfer efficiency, reducing that of air-diffusion systems and, in some cases,

30. The Newbridge Works of Midlothian County Council, employing Simcar aerators

31. An oxidation ditch employing extended aeration. This particular unit was the first in Britain to utilise the split-ditch principle, eliminating a separate settling tank

slightly enhancing the performance of mechanical aerators. Foam may be suppressed by anti-foam chemicals or by effluent sprays.

To obtain maximum future benefits from the oxidation units to be installed, the ease with which the treatment capability might be increased or automation incorporated should be considered at the design stage. As the rate of oxygen solution is proportional to the oxygen deficit of the liquid, energy will be wasted if the dissolved oxygen concentration is maintained at a level higher than that required by the micro-organisms for satisfactory oxidation (i.e. about 2 mg/l). By coupling dissolved oxygen meters to air control valves, the dissolved oxygen concentration in an air-diffusion plant can be controlled successfully. At one British plant producing a fully nitrified effluent, a full-scale, automatically controlled unit required 20 per cent less air than adjacent conventional units treating the same flow. With mechanical aerators, dissolved oxygen electrodes can be used to regulate aeration intensities by altering the degree of immersion of the aerators or by adjusting their rotational speed.

When the pollution load on a unit increases, the treatment capacity of a diffused-air plant can be raised by increasing the air supply at the expense of increased frictional losses through the diffusers, but at a saving in further construction costs. Similarly, if sufficiently large driving motors are initially provided on mechanical aeration units, increased pollution loads can be accommodated by increased speed or immersion of the aerators. At the lower loadings, motor efficiency would be reduced, but this would be a small premium for subsequent increased treatment capability; the associated lower power factor can be readily corrected.

Current trends

Despite the proven efficiency of activated-sludge treatment, a proposed treatment scheme may be too small to justify a conventional installation incorporating primary sedimentation, aeration, final settling tanks and associated activated-sludge return equipment. Instead, one of the extended aeration units mentioned earlier may be adopted to deal with unsettled sewage, incorporating aeration and final settling facilities within a single compact unit, or in the form of an oxidation ditch (photograph 31). When regular and frequent supervision of extended aeration units is not possible, the final effluent should be passed over grassland or sand, to prevent any spasmodic discharges of excessive suspended solids from impairing the quality of the effluent discharged to the watercourse.

In recent years, the trend towards increased aeration intensities has enabled shorter detention periods to be adopted at lower capital costs. Limitations resulting from the need to supply the increased volumes of air required in diffused air systems, on peripheral speed of mechanical aerators and on the satisfactory separation of the high concentrations of activated sludge necessary to utilise the increased oxygen have been resolved satisfactorily by process modifications and other British development work on aerator design and on the geometry and configuration of aeration and final settling tanks.

G. AINSWORTH

On leaving grammar school, Mr Ainsworth worked for six years in the laboratories of William Walker & Sons Ltd, leather manufacturers. During this period he gained an external BSc special degree in chemistry from the University of London. In 1948, although scheduled to take up a research fellowship at an American university, he accepted an appointment as chemist to the Bolton Corporation Sewage Department. In 1954, when the Bolton and District Joint Sewerage Board was formed, he became its deputy general manager and chief chemist and was associated with much of its development work. In 1957 he was appointed senior chemist to the Colne Valley Sewerage Board, and when that organisation became the West Hertfordshire Main Drainage Authority he was made its assistant general manager. In 1964 he was appointed general manager of the City of Manchester Rivers Department and is responsible for the sewage and trade effluent treatment of Britain's most highly industrialised area.

CHAPTER 7

Treatment of Industrial Effluents

BY A. B. WHEATLAND, BSc, AMInstWPC

Industry uses large quantities of water for purposes such as washing, classifying, conditioning and conveying materials and for general cleansing. During such uses, it takes into solution or suspension matter derived from raw materials, process chemicals, or products and must be treated before discharge to conform with the requirements of river authorities. In some areas of Britain, particularly in south-east England, the requirements are as severe as will be met anywhere in the world, since the rivers are small and many serve as a source of water for public supply. Experience over a decade or more has shown that the plant evolved has proved fully capable of meeting these exacting requirements reliably and economically.

An industry wishing to keep the total cost of its operations to a minimum must regard the use of water and its treatment and disposal after use as integral parts of the main productive activity. This approach can be applied with advantage both to the design of a new plant and to the operation of an existing one, although the best solution may not necessarily be the same in the two cases.

The size and cost of effluent treatment plant is clearly related to the volume of effluent to be treated per day and to the type and quantity of impurity it contains. Accordingly, any measure that will reduce the volume or strength or the impurity load will be helpful. In adopting such measures, careful consideration should be given to the possibility of reclaiming useful materials, including water, for re-use. Such materials in effluents represent a direct loss and also increase subsequent treatment costs. As examples, in the dairy industry, drainage of churns and prevention of spillage, leaks, etc, reduce losses of milk from reception depots and bottling plants from about one per cent of the volume of milk processed to about 0·25 per cent. In the electro-plating industry, the use of drag-out tanks and counter-current and spray rinsing techniques reduce water usage, losses of metal, and the cost of effluent treatment; furthermore, the use of ion-exchange produces high-quality, demineralised water for re-use. In the beet-sugar industry, recirculation of press water increases recovery of sugar and, with other similar measures, reduces the cost of effluent treatment by over 90 per cent.

In addition to the above measures, it is usually important to segregate process waters from uncontaminated yard and roof drainage and from cooling water. It is sometimes also beneficial to segregate strong liquors from weak ones and to treat individual process streams separately, when this facilitates regeneration of liquors or recovery of useful products.

An experienced consultant will consider the problem of effluent disposal at the earliest stage of planning any industrial development, since it may determine the possible location of a new plant. The following options may be open to him.

In Britain, industry has a legal right to dispose of waste waters to the public sewers, where this can be done without interfering with the efficiency of sewage treatment. A firm wishing to make a new discharge must first obtain permission from the local authority, and the authority's consent will usually include conditions designed to protect the sewers and the men who may have to enter them, to ensure continued efficient operation of the sewage treatment works, and to recover the additional treatment costs involved. Industry is encouraged to use such facilities and disposal in this way is often the most economical method.

Waste waters of high BOD may need partial treatment before discharge to a public sewer and other types of waste waters pre-treatment to neutralise excessive acidity or alkalinity, or to remove toxic metals and other substances which might interfere with any of the stages of sewage treatment. However, where treatment in admixture with sewage is possible, it will generally be the most satisfactory method of dealing with small volumes of effluent. British consultants and plant manufacturers take into account discharges of trade effluents in designing sewerage schemes.

If the volume of industrial effluent is too large for disposal to public sewers, or if suitable sewers do not exist, the industry must provide separate effluent treatment plant and dispose of the treated effluent direct to surface waters.

The nature of the unit processes needed to treat an effluent before discharge to surface waters depends on the type of waste and on the pattern of discharge. In broad terms, it is a matter of experience that the more uniform the flow of waste water and the more uniform its omposition, the easier it is to achieve effective and economical reatment, since the reserve capacity of the plant to cope with peak oads can be kept to a minimum when conditions are fairly steady. Thus, as is often the case, if the flow and composition of the waste vaters vary substantially, it will frequently be necessary to provide balancing' tanks to smooth out these fluctuations. If both acid and lkaline waste waters are discharged at different times, such tanks an be used to effect at least partial neutralisation, thus reducing the ost of chemicals for any subsequent adjustment of pH value.

Unit processes for the removal of suspended matter include screenng, sedimentation, filtration, centrifugation, flotation, and treatment n hydrocyclones. Chemical treatment may include adjustment of pH alue, precipitation of dissolved impurities, adsorption, ion-exchange, lectrolysis, oxidation, reduction, and solvent extraction. For wastes ontaining organic matter, the main processes will often be biohemical oxidation or reduction. In many of these processes, the mpurities are removed from the waste waters in the form of moist olids or sludge which may need further treatment before ultimate lisposal as described in other parts of this book.

Certain types of industrial waste water, particularly those from the ood industries, may be suitable for disposal by irrigation on land. This method of disposal is not as widely used in Britain as in some other countries, mainly because of climatic conditions, but British nanufacturers offer a wide range of spray equipment for irrigation.

There might be reasons for limiting the discharge of certain ersistent toxic chemicals by pipelines to the sea, but many types of ffluent can safely be disposed of in this way. British consulting ngineers have considerable experience in the siting and construction of pipelines for that purpose, using the most modern techniques.

Disposal in deep strata is employed only where geological condiions are such that waste waters can be disposed of without contamiating non-saline ground waters. It is used for a few industrial lischarges in Britain.

Treatment processes employed for different types of industrial effluent

An indication of the types of process adopted for treatment of articular industrial effluents in Britain is given in Table 2, though he list is by no means comprehensive. A full description of processes nd equipment would require an extensive treatise.

A few modern installations of interest are illustrated in the hotographs which follow.

32. A 70-foot (21 m) diameter thickener for effluent treatment plant at Imperial Smelting Corporation's zinc smelter at Avonmouth

33. Pipeline from Imperial Smelting Corporation's zinc smelter at Avonmouth, through which trade effluent is discharged to the strong tidal water of the Bristol Channel

34. Contact stabilisation plant treating trade effluent from a pharmaceutical plant at Windlesham, Surrey. The unobtrusive nature of the plant may be noted

Biological treatment processes

Biological processes are important for the treatment of industrial effluent as well as for treatment of domestic sewage. Aerobic processes such as those described in other chapters of this book are widely used for removal or destruction of organic matter and also in some cases for oxidation of inorganic constituents of effluents such as sulphide and ammonia.

The cost of introducing oxygen into water by artificial aeration can be as little as one penny (1 US cent) per kg, whereas oxygen added in a more reactive state, for example as sodium hypochlorite, is considerably more expensive (93 cents per kg). Aerobic biochemical oxidation processes are therefore much less costly than chemical oxidation processes, as far as running costs are concerned; they are often also very much more efficient for dilute wastes.

Anaerobic processes also find an application, particularly for treatment of strong organic wastes. The anaerobic process involves reduction of organic matter to methane which can be burned, and

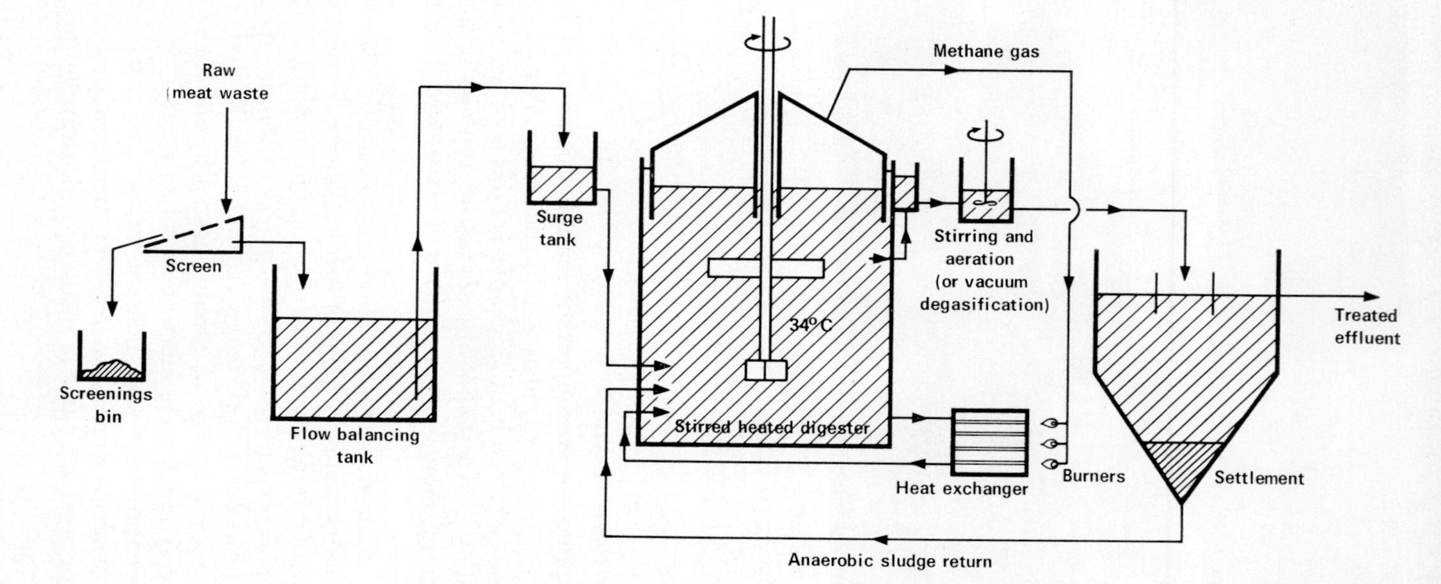

Fig. I. Flow diagram of plant for anaerobic digestion of meat waste

results in less surplus sludge than the equivalent aerobic process. No energy is needed to supply oxygen. The process is thus potentially cheaper than the aerobic one for strong wastes which yield sufficient combustible gas to warm the incoming flow to the optimum temperature of 34°C. In practice, this means that the process is mainly suitable for wastes containing not less than about 0·3 per cent organic matter (BOD $>$ 3000 mg/l). An example is meat waste waters (Fig. I).

Biological filtration plants are designed having regard principally to the rate of flow and strength of the waste waters and the quality of effluent required. For example, it will be found that a typical design loading for a plant in Britain required to treat domestic sewage at rates up to three times the average dry-weather flow to give effluent having a BOD not exceeding 20 mg/l is 0·09 kg BOD/m^3 d. With settled sewage of BOD 250 mg/l, this corresponds to a hydraulic loading of 0·36 m^3/m^3 d. A similar design loading is satisfactory for industrial effluents of generally similar character and strength under the conditions prevailing in Britain. Some types of industrial

35. *One of the first biological filtration plants installed anywhere specifically for the treatment of trade effluent containing formaldehyde and phenols*

36. External view of a typical plant for treatment of chrome and cyanide liquors and rins waters from electro-plating processes at a large manufacturing works

37. Indicator panel and flow diagram in a typical plant for the treatment of chrome an cyanide liquors and rinse water from electro-plating processes

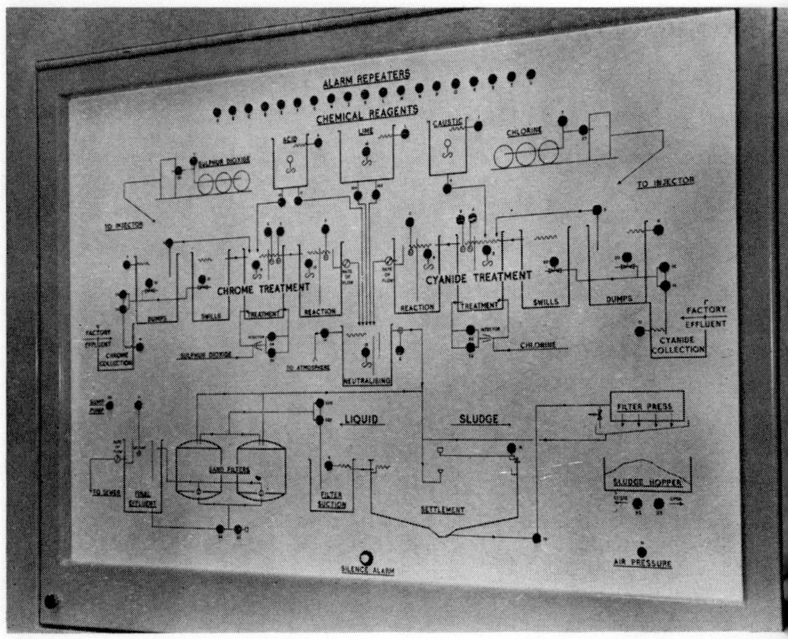

effluents are more easily and some much less easily treated than domestic sewage. Experience over many years has provided an adequate basis for design in the case of most common types of effluents, and well-established techniques are available for the assessment of the treatability of those of unusual composition. In the case of waste waters having a high BOD, provision is usually necessary for recirculation of effluent.

Where partial treatment is adequate (say, 50–70 per cent removal of BOD), or for the first stage of a two-stage plant, treatment is possible at very high loadings on either coarse rock or plastics sheet medium. Examples have been given in a previous chapter.

Activated sludge

The rate of flow and the oxygen demand of the waste water and the quality of effluent required are clearly important factors to be taken into account in designing an activated-sludge plant. Also important are parameters such as volumetric loading and sludge loading. When a nitrified effluent is required, the relative proportions of carbonaceous matter and nitrogenous substances must also be taken into account. Recent research has shown that, for a given concentration of activated sludge and for a given waste water at a particular temperature, there is a minimum period of aeration below which no nitrification is obtained under equilibrium conditions and above which substantially complete nitrification occurs.

Formulae defining this minimum period in terms of operating variables, such as the BOD of the waste water, the concentration of sludge maintained, the temperature and pH value of the mixture of sludge and waste water, have been evolved which can be used as a basis for design.

When considering the treatment of industrial effluent subject to considerably greater variations in rate of flow, composition and strength than domestic sewage, the conditions of mixing in the aeration tanks may be of considerable importance. Often it is preferable to have complete mixing. This system has the advantage that if potentially toxic or inhibitory substances are present in the trade waste, they are rapidly diluted by the total volume of mixed liquor and reduced to a concentration below the toxic threshold. In the aeration channel of the other type of plant, there is little longitudinal mixing, dilution is less rapid, and the effects of toxic substances much greater.

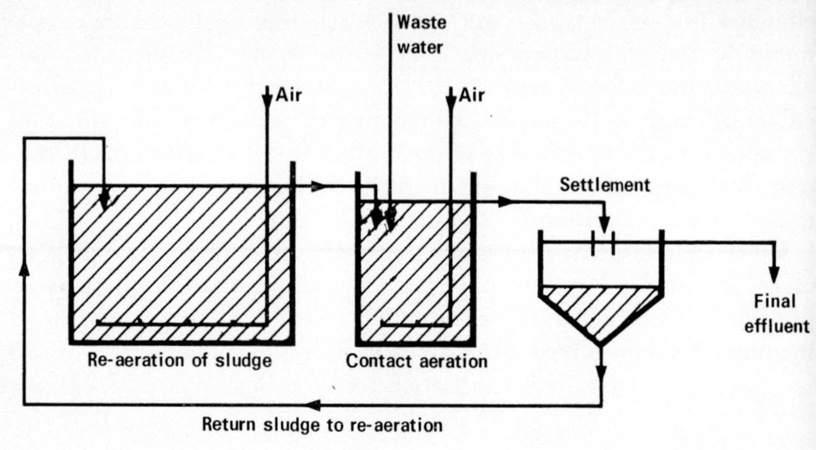

Fig. II. Flow diagram of contact stabilisation process

A contact stabilisation plant (Fig. II) has some of the advantages of the completely mixed system; an additional advantage which may sometimes be important is that there is a large reserve of micro-organisms which are not immediately affected by an accidental shock discharge and may be used to restore the efficiency of the process when the toxic effluent has been removed.

Treatability studies

The treatability of an industrial effluent by a biological process can be assessed in several ways. In principle, if the waste waters have a measurable biochemical oxygen demand, it will be possible to satisfy that demand by biological treatment, provided the waste water can be sufficiently diluted to reduce the concentration of toxic substances below the toxic threshold as in the BOD test, and provided any deficiency of nutrients such as nitrogen, phosphorus and potassium is made good.

The treatability can also be assessed by bringing a sample of the waste water into contact with activated sludge in the presence of dissolved oxygen and measuring the change in the rate of respiration of the sludge. An effluent containing substances easily assimilated by micro-organisms will cause the rate of respiration of the sludge to increase significantly, whereas waste waters containing substances inhibitory to bacteria or resistant to degradation will result in either a decrease in the rate of respiration or little change.

The presence of substances which are inhibitory to specific rganisms, such as those which bring about the oxidation of ammonia, an be established by comparative tests in which a sample of nitrify-ng activated sludge is aerated with a mixture of effluent and lomestic sewage and, as a control, with a mixture of tap water nd domestic sewage. After two hours, bacterial action is stopped by cidifying the samples and the concentration of oxidised nitrogen ormed is measured. An increase in concentration of oxidised nitrogen ower by more than 25 per cent than that in the control is significant nd indicates that the waste water contains substances having an mmediate inhibitory effect on nitrification. For those substances aving a cumulative effect, longer-term tests are necessary.

Some useful information on the treatability of waste waters can be btained from experiments in respirometers but it is more satisfactory o operate small continuous-flow plants which more closely simulate he treatment process and provide conditions for selection or adapta-ion of the micro-organisms to occur. For a preliminary assessment of iological filtration, rolling tubes are convenient (Fig. III), but as a asis for design of full-scale plant it is generally more satisfactory to arry out tests with pilot-scale filters unless there is already experience vith similar effluent in full-scale plant.

For preliminary studies of the possibility of treating particular vaste waters by the activated-sludge process, simple fill-and-draw lants may be manually or automatically operated (Fig. IV), but as vith filters, it is best to operate small continuous-flow plants on site efore designing full-scale plant for treating any new type of waste vater. In such studies it is usual to vary the period of retention of the vaste water and the concentration of activated-sludge solids main-ained in the mixed liquor. Among the parameters which it will be mportant to measure will be the BOD and the suspended-solids

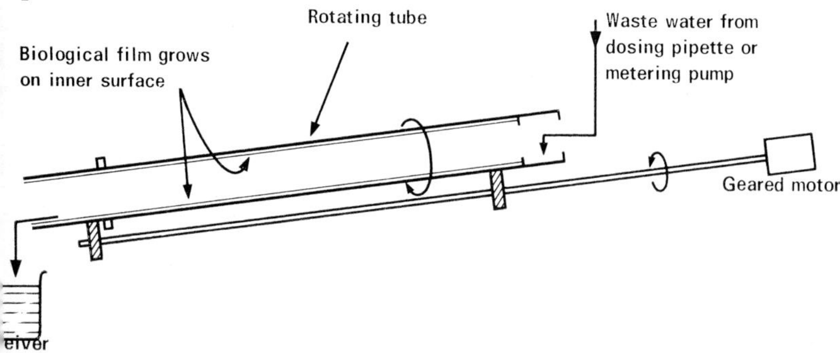

ig. III. Rolling tube for preliminary assessment of biological treatment

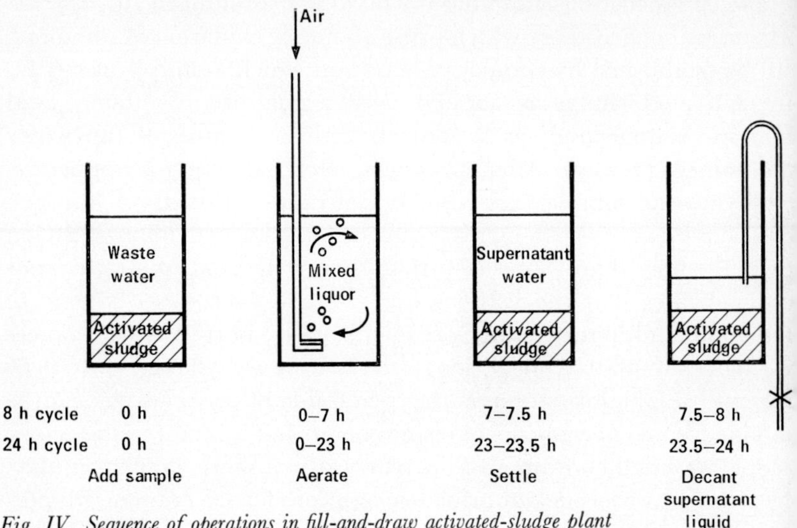

Fig. IV. Sequence of operations in fill-and-draw activated-sludge plant

content of the raw waste and the treated effluent, the rate of respiration of sludge, the settlement properties of the sludge and the rate of production of surplus activated-sludge solids. On the basis of such studies in pilot plant, activated sludge plants have been successfully designed for treatment of effluents from a number of industries for which there was no established method of effluent treatment a few years ago. These industries include synthetic fibre and rubber latex production, bleaching and dyeing, pharmaceuticals, etc. A simple flow diagram for one plant is shown in Fig. V.

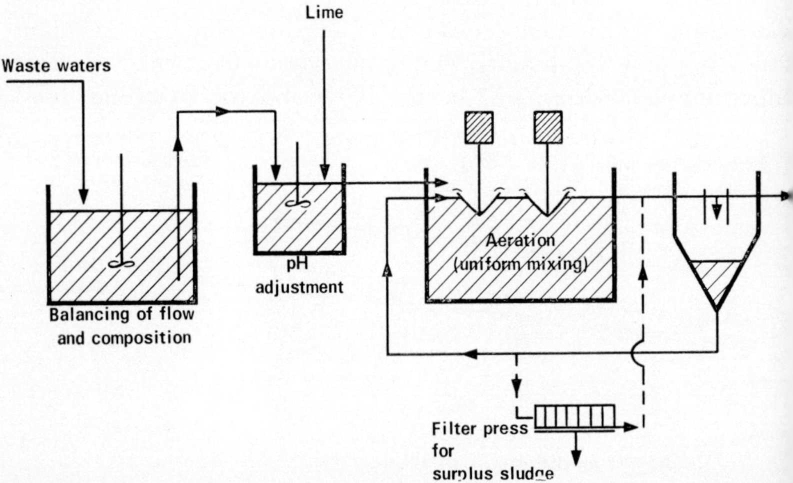

Fig. V. Flow diagram of plant for treatment of 'weak' effluent (BOD 5000 mg/l) from manufacture of fine pharmaceutical chemicals

TABLE 2. Types of plant installed for treatment of industrial effluents

Key

1 Biological filtration
2 Activated sludge
3 Anaerobic digestion
4 Chemical treatment
5 Electrolysis
6 Physical processes including evaporation, incineration, spray irrigation

Source	Type of plant (*see key*)	Source	Type of plant (*see key*)
Abattoir	1 2 3	Meat processing	1 2 3
Adhesives manufacture	2 4	Milk processing	1 2 6
Anodising	4	Mineral processing	4 5 6
Antibiotics production	1	Mining	4 6
Automobile manufacture	4 6	Nuclear power generation	4
Beet-sugar	1	Oil refinery	4 6
Bleaching & dyeing	1 2 4	Paper industry	1 2 4 6
Brewing	1 2	Pea processing	1 2 6
Canning	1 2 6	Pharmaceutical industry	1 2 6
Cattle sheds	1 2	Phosphatising	4
Chemical industry	4	Pickling metals	4 5
Chocolate manufacture	1 2	Piggeries	1 2
Cider production	1	Plastics emulsion manufacture	4
Coffee processing	1	Plastics manufacture	1 2 6
Coke ovens	1 2	Poultry processing	1 2 6
Cotton kiering	6	Quarrying	4 6
Dairy	1 2 4 6	Rendering offal	1
Distillery	1 6	Retting flax	1 2
Dyeworks	1 2	Rubber latex production	2 4
Egg processing	1 6	Starch production	1 2 6
Electro-plating	4 5	Stone working	4 6
Engineering	4 6	Synthetic fibres	1 2
Engraving	4	Tall oil processing	2
Farming	1 2 6	Tanning	1 2 4
Fellmongering	1 2	Textile manufacture	1 2 6
Felt manufacture	4	Transistor manufacture	4
Fermentation	1 2 6	Vegetable processing	1 2 6
Fish meal production	6	Vegetable washing	6
Gas wash	2	Vulcanised fibre production	4
Glass working	6	Wire drawing	4 5
Grain washing	1	Wool scouring	4 6
Iron & steel industry	4 6	Yeast manufacture	1 2 6
Malting	1 2		

A. B. WHEATLAND

Mr Wheatland joined the Water Pollution Research Laboratory from school and, after study at evening classes, graduated as an external student of London University in 1948. He is now in charge of the biochemical engineering section of the Laboratory and is concerned with research, sponsored investigations, and advisory work on trade effluent and other problems. He has worked on the treatment and disposal of waste waters from plastics and paper manufacture, beet-sugar production, and metal finishing, and has carried out research in connection with treatment of potable water, recharge of aquifers, and the effects of pollution on estuaries and coastal waters.

CHAPTER 8

Tertiary Treatment

BY G. A. TRUESDALE, BSc, FInstWPC, MIPHE

Conventional biological treatment, such as has been described in earlier chapters, can produce an effluent of '30/20 standard', and often better, but for reliable production of higher quality effluent, such as is nowadays often required, a tertiary or 'polishing' stage of treatment is necessary. Polishing processes rely mainly on flocculation, sedimentation or filtration of much of the residual suspended solids which have escaped sedimentation in humus tanks or in the final separation tanks of activated-sludge plants. The BOD associated with the solid is also removed, and some of the processes provide further biological purification. It is important to stress that 'polishing' procedures are suitable only for dealing with well-oxidised effluents and that no process—with the possible exception of grassland treatment—will operate effectively at works where biological treatment is inadequate.

Several methods are now available. These include some 'natural' processes such as surface irrigation over grass plots, slow sand filtration, or retention in lagoons, while others of a more mechanical nature include microstraining, rapid sand filtration, and flocculation and straining in upward-flow gravel beds.

Irrigation over grassland

This treatment, applied to sewage effluents, should not be confused with the old-fashioned land treatment of sewage. It is particularly suited to small works in rural areas where land is available; it is a cheap and effective method of removing suspended solids and bacteria. In this method, effluent is distributed over the land from a system of channels and, after flowing across the surface, is collected in a second system of channels. The land should be well graded and, to avoid erosion of channels, should have a gentle slope of about 1 in 100. With specially prepared plots, it is usual to sow a special mixture of deep-rooted grasses, though it appears that, within a few years, a more natural flora is restored; indeed, plots which have not been specially seeded appear to perform quite satisfactorily. Occasional cutting to keep the growth from becoming too rank is required, and it is necessary to rest a plot periodically—after two to three months' continuous operation—to allow it to dry out, the flow being transferred to a second area. Grass plots may be used for several years before it becomes necessary to remove the accumulated solids, the period of use depending on the solids content and rate of treatment of the liquid.

Lagoons

Storage of liquid in lagoons to allow removal of suspended matter by deposition, and other organic matter by biological oxidation, is another method that has been used for improving effluents.

Shallow (3 ft 3 in [one metre] deep), short-retention (three to four days) lagoons have been found to be only moderately effective, reducing the suspended solids content by no more than 50 per cent. Also, they tend to suffer from rising and floating sludge, and algal growths may cause a seasonal increase in solids in the effluent discharged. Passing the effluent through several lagoons in series reduces the possibility of solids from rising sludge appearing in the final discharge.

Deeper lagoons, however, operating with retention periods of up to 17 days, can remove about 80 per cent of solids from sewage works effluents, with corresponding bacterial reductions of 99 per cent. Again, there may be seasonal difficulties from algal growths.

Slow sand filters

These are similar to the slow sand filters which are still used at some water works and are suitable only for small works. In its simplest form, the slow sand filter is a bed of sand 1 to $2\frac{1}{2}$ ft (300 to 750 mm) deep, resting on a layer of coarser material which in turn rests on a system of underdrains. Settled effluent is applied to the filter at a rate of up to 10 ft^3/ft^2 d (3 m^3/m^2 d) and percolates through the bed of sand until the head loss becomes excessive, a level usually determined by the height of the retaining wall. The filter is then drained—the flow being diverted to a second bed—and the surface layer of sand with the accumulated solids is removed. From time to time the loss of sand is made good by adding more sand. The process is reasonably effective in removing suspended matter, but it is expensive.

Rapid gravity sand filters

Rapid gravity sand filters are essentially similar to those commonly used in water works, and they are well established at a few of the larger sewage works where skilled maintenance and regular supervision can be provided. A filter consists of a layer of graded sand, 3–5 ft (1–1·5 m) deep, supported on layers of graded gravel resting in turn on a special filter floor in which is set a series of nozzles for collecting the filtrate, the whole being held in a suitable casing. The underdrainage system is provided with facilities for backwashing and air scouring. The liquid to be treated flows downwards through the sand at a controlled rate of 300–500 ft^3/ft^2 d (90–170 m^3/m^2 d)—a flow very much greater than for slow sand filters. As filtration proceeds, the loss of head increases because of the accumulation of suspended matter and, at a certain time each day, or when the head loss has increased to about 10 feet (3 m), the filter is backwashed using filtered effluent, air-scour being employed to assist the separation of sludge from sand. The backwashings, amounting to some 2½ per cent of the volume filtered, are returned to the inlet of the works. The process is equally effective with either humus-tank or activated-sludge effluents.

Immedium upward-flow sand filter

This filter (Fig. VI) is a comparatively recent innovation in filtration practice; its application as a process for polishing sewage works effluents has been developed in Britain. The filter bears a superficial resemblance to the rapid gravity filter, but the liquid flows upward through the sand. Backwashing grades the sand with the coarser material at the bottom and the finer material at the top. Thus, during upward filtration, the effluent meets progressively finer material and the finest sand is protected from blocking with the coarser solids. Solids removed by filtration are thus distributed throughout the sand rather than being concentrated near the surface as in downward filtration. Higher rates of treatment (up to 1300 ft^3/ft^2 d [400 m^3/m^2 d]), and longer filtration runs are thereby obtained. To prevent expansion of the sand at these high rates, it is held down by a metal grid, consisting of parallel bars spaced 4–6 in (100–150 mm) apart, just below the surface of the bed.

The filter consists essentially of a suitable casing containing a 5–7 ft (1·5–2 m) deep bed of sand, supported on a 1 ft (300 mm) depth of graded gravel, resting in turn on a special floor fitted with a liquid distribution system. Cleaning of the bed is achieved by an

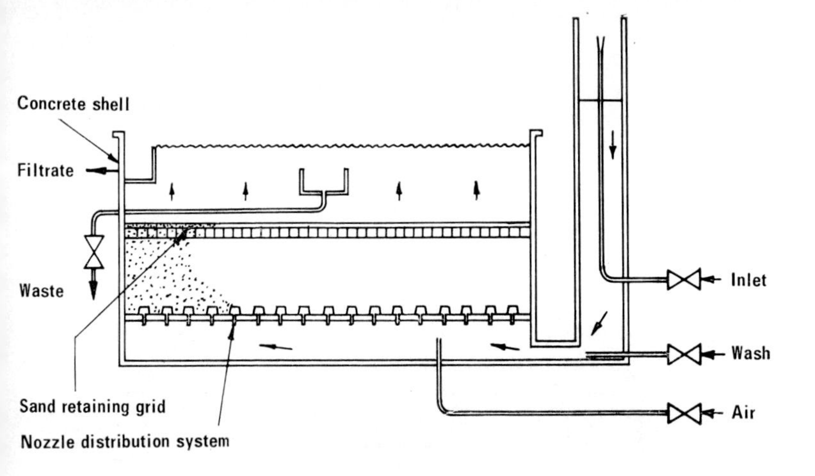

OPEN TYPE

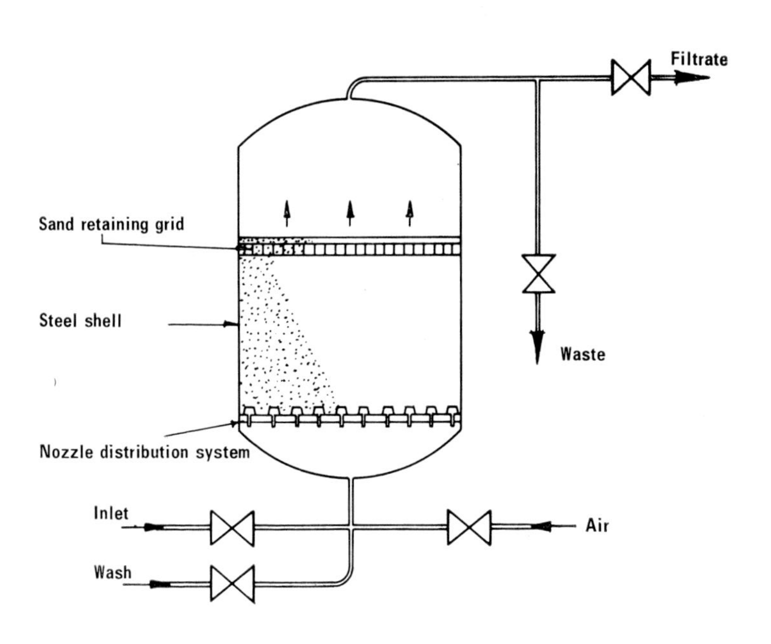

PRESSURE TYPE

Fig. VI. Immedium upward-flow sand filter

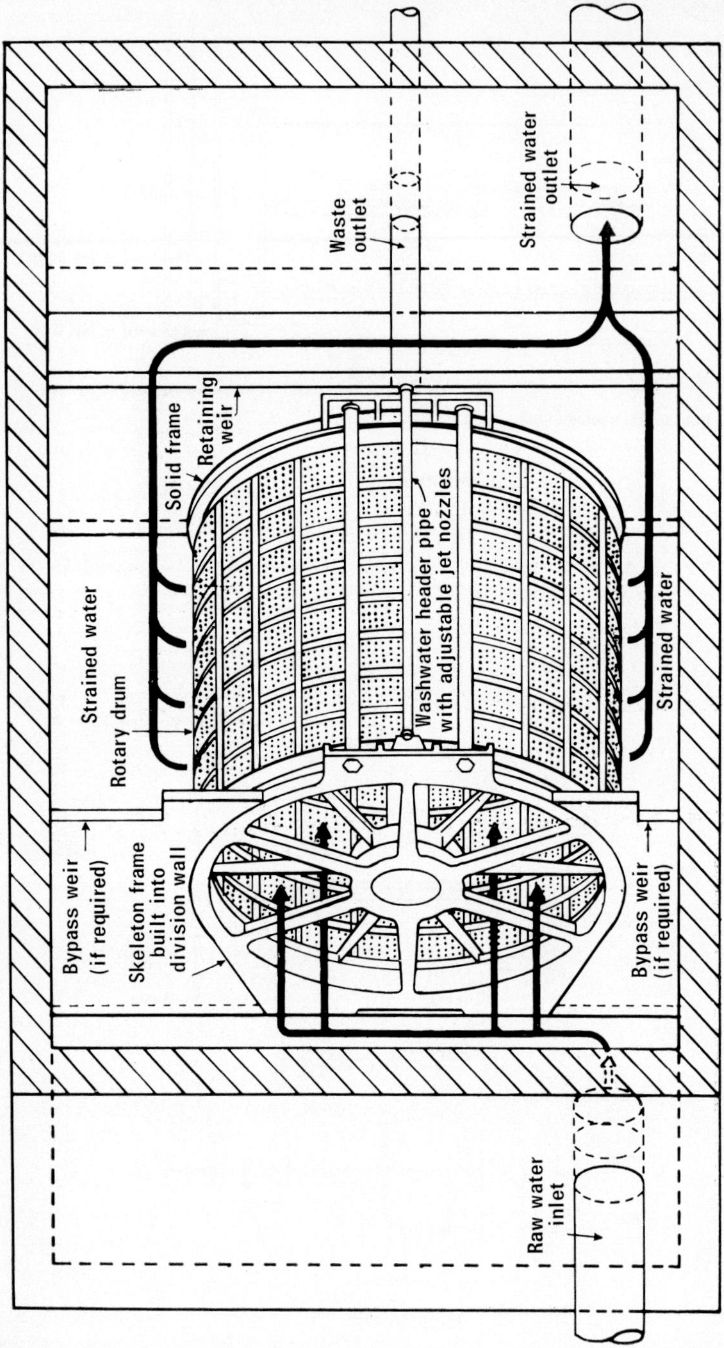

Fig. VII. The microstrainer

ncrease in the upward flow of unfiltered effluent to about 2500 ft^3/ft^2 d 800 m^3/m^2 d) to cause the bed to expand through the restraining grid. The solid matter separated from the sand, together with the backwash water which represents some 2–3 per cent of volume filtered, is passed out of the unit. The filter may be cleaned automatically when the pressure head against pumping reaches a certain value.

In extensive pilot-scale trials, the performance of this filter has been found to be similar to that of the conventional sand filter but the higher rate of treatment saves space and cost. The first full-scale application of this process in the sewage-treatment field will be at Luton sewage works, in Bedfordshire, England, where a plant treating a flow of 10 mgd ($4 \cdot 5 \times 10^4 m^3/d$) will be in operation in 1969.

Microstrainers

The microstrainer (Fig. VII) was developed originally by Glenfield and Kennedy Ltd for the treatment of raw water supplies, but has since proved to be a suitable device for improving the quality of sewage effluents. It consists of a horizontally mounted drum, closed at one end, the cylindrical surface of which is formed from a special stainless-steel fabric, which rotates slowly in a tank with two compartments so that effluent enters the open end of the drum and flows out through the filtering fabric. As the drum rotates, solids collect on the inner surface and are continuously removed from the fabric by strained effluent, which is pumped under pressure through a row of jets fitted on top of the machine and extending the full width of the drum. The wash-water containing the removed solids and amounting to about 2 per cent of the volume treated is returned to the works' inlet. Biological growths on the fabric are suppressed by ultra-violet light from a lamp mounted alongside the wash-water jets.

Microstrainers have been used successfully for several years at a number of large- and medium-sized works for polishing humus-tank effluents. For this application, Mark I fabric (nominal apertures 35 μm) is fitted. Recently, the method has been applied to effluents from activated-sludge plants, for which purpose a finer material (Mark 0) with nominal apertures of 23 μm is used.

Trials by the Water Pollution Research Laboratory at a sewage works during 1962–63 showed that the microstrainer could be used successfully to treat well-oxidised effluent direct from percolating filters, thus dispensing with the humus tank. Over a period of 11 months, the microstrainer was found to be consistently superior to a humus tank operated as a control and having a retention period of three hours. The microstrainer removed on average 73 per cent of suspended solids compared with 52 per cent by the humus tank.

The gravel-bed clarifier

This device is a means of clarifying effluent by upward flow through a shallow bed of gravel. In this system, effluent is passed upwards through a 6 in (150 mm) layer of $\frac{1}{4}$ in (6 mm) gravel supported on a perforated floor in a suitable tank, at a rate of up to 75 ft^3/ft^2 d ($23\ m^3/m^2$ d). Solids which accumulate in, above, and below the gravel are removed by backwashing, that is, by lowering the water level by running off effluent from below the gravel bed; the process may be completed by washing the drained bed with a jet of water or effluent.

Comparison of processes

If a suitably chosen polishing process is applied to a well-oxidised secondary effluent, it should normally be possible to reach a 10/10 standard (10 mg/l suspended solids, 10 mg/l BOD) of quality, and with special care a 7/7 standard. Performance data for the methods described are compared in Table 3.

TABLE 3 Comparison of tertiary treatment processes

Method	Rate of treatment (maximum) (ft^3/ft^2 d)†	Performance (per cent removal) Suspended solids	BOD	Bacteria (coli-aerogenes)	Approximate cost* (pence**/m^3)
Grass plots	2·8***	70	50	90	0·09
Lagoons (short retention)	1·6	40	40	70	0·07
Lagoons (long retention)	1·0	80	65	99	0·13
Slow sand filters	10***	60	40	50	0·46
Rapid gravity sand filters	560	80	60	30	0·16
High-rate upward-flow sand filters	1300	70	55	25	0·11
Microstrainers	1300****	70	40	15	0·18
Upward-flow gravel-bed clarifiers	75	50	30	25	0·15

† 1 ft^3/ft^2 d = 0·3 m^3/m^2 d.
* Capital and running costs. Interest charge 6 per cent per annum.
** 1 penny = 1 US cent.
*** Calculated on area in use at any one time.
**** Flow per unit area of fabric.

Rates of treatment differ widely, being greatest with rapid sand filters and microstrainers. These methods are generally used at medium-sized and larger works. Although approximate average costs of treatment are indicated in the table, with each of these processes the capital cost for treatment of a given volume of liquid decreases with increasing size of unit.

Of the three methods recommended for smaller works—grass plots, slow sand filters and lagoons—the first two appear to be consistently effective in achieving a substantial clarification. Lagoons may be unreliable at certain times of the year because of the tendency for sludge to rise to the surface, and for planktonic algae to be produced which may pass out of the lagoon in the effluent. The gravel-bed clarifier, in view of its performance and the ease with which it may be installed in existing final sedimentation tanks, is a particularly attractive process for the smaller and possibly medium-sized works. Costs of these tertiary treatment processes are relatively low, adding a further 10 to 20 per cent to the cost of primary and secondary treatment.

Water reclamation

Following the evolution of polishing processes, it was but a short step to the realisation that, by using comparatively simple procedures, polished sewage effluents could be purified still further at reasonably modest cost to produce water that would be suitable for a wide range of industrial purposes; and that this approach could, in some areas, afford a useful means of reducing the pressure of demand on supplies of potable water.

It was demonstrated by the Water Pollution Research Laboratory that, by treating a polished effluent first by aerating it in a foaming column to remove detergents, and then by standard water treatment techniques of coagulation, sedimentation, filtration and chlorination, a water can be produced which is clear, colourless, sterile, free of suspended matter and low in phosphate content (about 1 mgP/l). The cost of this treatment is dependent, to a large extent, on the amounts of chemicals employed, but is thought likely to be about twopence per cubic metre.

A water of a similar quality to that described above may also be obtained by employing a process embodying ozonation and known as the MD (Micellisation-Demicellisation) process. The system involves, first, treatment by a microstrainer fitted with a Mark 0 fabric to remove particles down to about one micron in diameter, then treatment by ozone, followed by rapid filtration through sand. This process has been examined during the operation of a pilot plant treating 10 m3/h of sewage effluent at a works in the south-east of England.

Compared with the original water supply, the final product from each of these processes still contains high concentrations of organic matter, some 340 mg/l more dissolved solids, and nitrates. Increasing attention is being given to the possibilities of recovering water of even higher quality at an acceptable cost, perhaps even suitable for potable use. A wide range of processes is available which might find an application for this purpose. These include adsorption on carbon, which is an effective means of removing certain types of organic matter; ion-exchange and electrodialysis, which are particularly suitable for removal of inorganic impurities; and reverse osmosis, which, though still requiring further development for practical application, is effective in eliminating a wide range of both inorganic and organic impurities.

G. A. TRUESDALE

Mr Truesdale joined the Water Pollution Research Laboratory in 1947 and worked there for 21 years before taking up his present appointment as chemical inspector at the Ministry of Housing and Local Government. At the Laboratory he was concerned with pollution by synthetic detergents, the development of standard procedures for assessing biodegradability, aeration, biological filtration, and an appraisement of various methods for further improving final effluents and for the reclamation of water from effluents. He is author of some 45 published technical papers on these and related topics.

CHAPTER 9

Sludge Technology

BY J. D. SWANWICK, BSc, PhD, ARIC, MInstWPC

The treatment of domestic and most industrial waste waters results in the production of sludges consisting of insoluble and frequently highly putrescible solids, retaining up to 100 times their weight of water. These sludges must be disposed of and the operations involved generally account for a substantial proportion of the total expenditure on waste water treatment; in some cases they also give rise to particularly difficult operational problems. It would not be feasible, nor indeed appropriate in this short study, to attempt a complete survey of all the many processes and combinations of processes that have been employed by British engineers for the treatment and disposal of sludge. The main methods are indicated in Table 4.

TABLE 4 Some processes used for treatment, dewatering and disposal of sludge

Process	Method
Thickening	(i) Gentle stirring
	(ii) Floatation
Digestion	(i) Aerobic
	(ii) Anaerobic
Heat treatment	
Composting with domestic refuse	
Elutriation	
Chemical conditioning	
Dewatering	(i) Drying beds
	(ii) Roto-plug
	(iii) Vacuum filtration
	(iv) Filter pressing
	(v) Centrifugation
Heat drying and combustion processes	(i) Heat drying
	(ii) Incineration
	(a) Multiple-hearth
	(b) Fluidised-bed
	(iii) Wet-air oxidation
Final disposal	(i) Land fill
	(ii) Use as agricultural fertiliser
	(iii) Discharge to sea

The most convenient and economical method of disposal at any given site depends on many factors, not least the location of the works. At most coastal towns, sewage and sometimes industrial waste waters are discharged to the sea through submerged pipelines without separation of sludge, and in such cases no special problem arises. Some of the larger towns and cities on estuaries convey sludge, usually after treatment, to dumping areas at sea, using special vessels. At inland towns, conveyance of sludge to the sea would be prohibitively costly and it is mainly disposed of on land, frequently after treatment and dewatering. Among traditional methods of disposal of sludge from sewage treatment, the most important one is by anaerobic digestion, followed by drying on open beds and disposal of the dried sludge on agricultural land. Dewatering by mechanical means, especially after addition of chemicals to condition the sludge, is quite commonly employed; for example, as an alternative to the use of drying beds. An increasing number of applications is being found for a variety of other processes, including heat treatment, wet oxidation, centrifugation and incineration.

The continuing quest for new, improved methods to deal with sludges emphasises the importance of the problem of treatment and disposal, and is a reminder that no single method is without some limitations. Great care and wide experience are essential in choosing the best procedure for each application.

It is vital to regard sludge treatment and disposal not as independent operations to be added on to the main scheme of purification, but as an integral and interdependent part of the whole. The reason is, first, that, following extensive research, it has been recognised that the characteristics of sludges depend closely and often predictably upon the conditions in the unit processes from which they arise; and, secondly, that the liquid separated from them during treatment generally contains considerable concentrations of polluting matter, which is commonly dealt with by recycling the liquors for treatment in the main stages of purification. Perhaps the most important advance in the last decade, strengthening British design practice, has been the introduction of quantitative methods for assessing the degree of difficulty of withdrawing water from sludges from measurements of their specific resistance to filtration, and for predicting the performance of full-scale dewatering plant from the results of such measurements. This approach, which has been widely adopted in other countries, has been greatly assisted by the development at the Water Pollution Research Laboratory of a simple

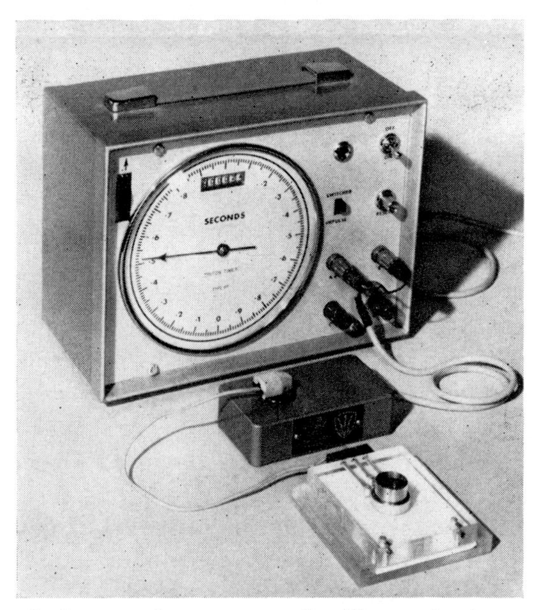

38. Apparatus for measurement of capillary suction time

automatic instrument utilising the capillary suction pressure of a filter paper to withdraw water from the sample (photograph 38). The instrument permits very rapid measurements to be made, thus facilitating process control, especially, for example, the control of the rate of addition of conditioning chemicals to correspond with changes in sludge characteristics. An additional advantage for the plant designer is that, because of the small volume of sludge required (5 ml), samples can be readily transported by air from remote locations where there are no laboratory facilities.

Use of these new techniques has enabled considerable increases in the throughput of particular installations to be achieved.

TREATMENT PROCESSES

Anaerobic digestion

Anaerobic digestion, as a separate treatment process for sewage sludge, was first developed in Britain and has been widely adopted elsewhere. The process serves at least half the population of Britain at the present time and British firms have installed many plants throughout the world. The designs adopted provide for two of the most important features—adequate mixing and control of scum—in a simple way (Fig. VIII) and appear to be at least as effective as more complex and expensive designs used elsewhere. Advantages of the process assume relative importance dependent on the particular application, but in most cases the removal of all offensive smell is very important. Nearly one-half of the solids are converted into gas and the bulk of the sludge can be reduced by approximately 50 per cent by the withdrawal of water which separates when digested sludge is allowed to cool under quiescent conditions (secondary digestion). The oxygen demand of soluble constituents in liquor removed from digested sludge is low (about 100 mg/l BOD) and therefore imposes only a small additional load on the sewage treatment process. Many pathogenic organisms are destroyed and up to 90 per cent of the grease is gasified. Sludge is therefore converted into a product readily acceptable on agricultural land and because over one half of the available nitrogen is in solution, liquid digested sludge provides a particularly useful nitrogenous fertiliser.

Anaerobic digestion may also improve the filtrability of sludge; this, however, may not apply in cases where a particularly difficult secondary sludge is involved, when the filtrability may deteriorate and anaerobic digestion also increases the coagulant demand of the sludge solids. Even in these cases, however, the subsequent dewatering and disposal stages are facilitated by the reduction in the quantity of solids. The gas produced (70 per cent methane) may be used as a source of fuel and for the production of electricity.

The methane-producing bacteria, in particular, are susceptible to inhibition by a variety of constituents of waste waters. Certain materials, such as chlorinated hydrocarbons, are particularly toxic and, as a result of the experience gained in dealing with a number of cases of inhibition during recent years, convenient small-scale methods have been developed for monitoring sludges to provide an early warning of the presence of toxic constituents likely to affect the main plant. Should toxicity be detected, techniques are also available for locating the source of toxic discharges. Synthetic detergents discharged mainly from domestic sources have been known to inhibit digesters at certain British plants. Confirmation of this cause of inhibition has recently been made possible by neutralising the detergents with stearine amine, and this material also provides an economic method of correcting full-scale digesters inhibited by detergents.

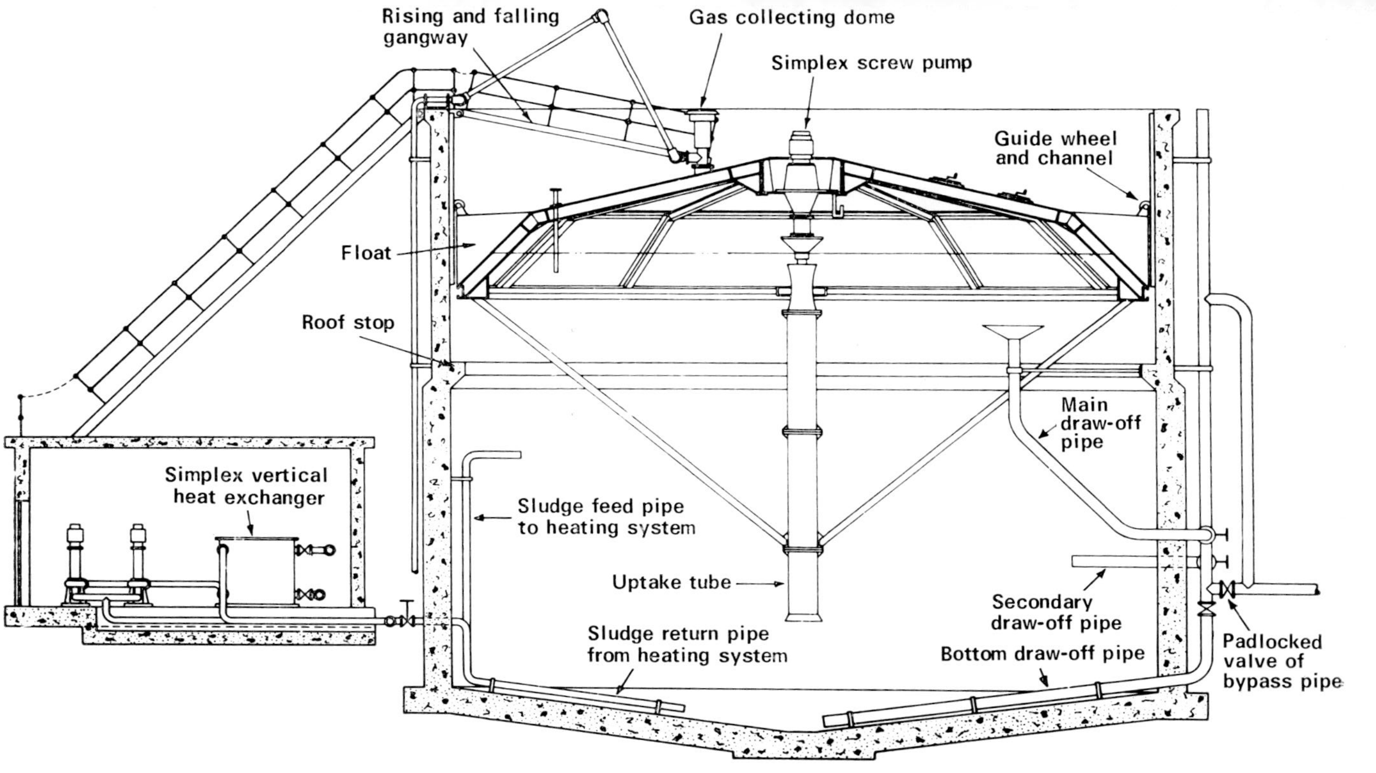

Fig. VIII. Cross-section of a heated digestion tank

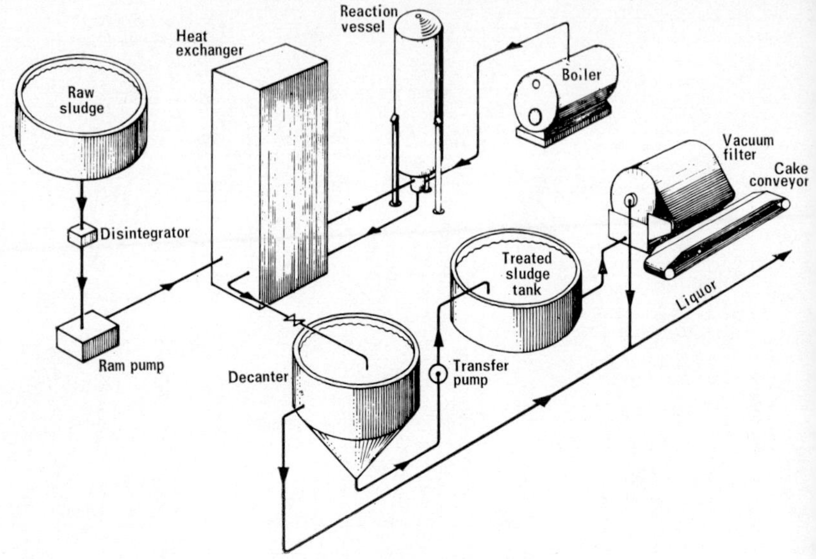

Fig. IX. Flow diagram of a heat treatment plant

Heat treatment

Heat treatment, involving the heating of sludge under pressure to temperatures of 180–200°C, has been used in Britain for over 30 years to improve the filtrability of sludge. The effect is similar to the best which can be achieved by chemical conditioning and the process can be operated on a continuous basis. Recent advances in the engineering field have stimulated interest in the method, both in Britain and in many other countries (Fig. IX). The process is applicable to all types of sewage sludges, including intractable ones which do not respond to chemical conditioning or anaerobic digestion. A proportion of the sludge solids is converted into soluble matter, giving a liquid effluent which may either be returned for treatment with the main sewage flow, or dealt with by other means. British practice favours filter pressing of the treated sludge, giving cakes with a dry solids content of up to 60 per cent. As a preliminary to land disposal, sterilisation of the sludge is an important advantage.

Wet-air oxidation

Wet-air oxidation, in which up to about 70 per cent of the oxygen demand is removed by reaction with atmospheric oxygen at temperatures of about 250°C and pressures of about 1400 lb/in^2

(110 kg/cm²), is used at a few plants in the USA for the treatment of sewage sludge, but interest in this method appears to have been concentrated recently on operation under less rigorous oxidising conditions, similar to those of the British heat treatment process. The conditioning effect is similar to that in the British process and the effluent also appears to have similar characteristics.

Freezing

Freezing was developed and is used in Britain for treating waterworks sludge, providing excellent irreversible conditioning of the solids with no effluent problem. The high power requirements have so far restricted its application in Britain, where the cost of power is high, but the availability of cheap power elsewhere could make the process highly attractive, especially since the liquor separated from the conditioned solids is likely to be much less polluting than that from other treatment processes.

Dewatering

The choice of methods of dewatering is largely influenced by the type of sludge to be processed. Sludge characteristics—as already indicated—depend upon operational conditions in various unit processes and may also be influenced by national circumstances. For example, sewage sludges in the USA appear to have a higher content of fibrous matter than those in Britain, and a higher proportion of sewage works in the USA have no secondary treatment of sewage. This may explain why vacuum filtration in particular appears to have been more successful in the USA than in Britain.

Drying beds

Drying beds constitute the traditional method of dewatering sewage sludge in Britain, but their continued application on the present scale is undoubtedly due to British developments in mechanisation, which permit automatic coultering of drying sludge to promote cracking, cake lifting and the re-sanding and levelling of beds with very small labour requirements (photograph 39). Recent research at the Water Pollution Research Laboratory has provided a fundamental understanding of the main factors influencing the method and has enabled the prediction of drying bed performance and requirements. The main disadvantage of beds is their dependence upon natural evaporation, which in some countries is subject to extreme vagaries of climate.

39. A 'part cut' type of sludge lifter

Mechanical methods

Such methods are independent of natural drying, depending instead generally on preliminary improvement of filtration characteristics by chemical conditioning. For many years, lime has been used for this purpose, either alone or in conjunction with salts of multi-valent metals. Although lime is relatively inexpensive, it has the disadvantage of adding appreciably to the bulk and dry weight of material for final disposal, and if the sludge is ultimately incinerated, some of the heat liberated by combustion of organic matter will be unavoidably absorbed during dissociation of the lime and cannot therefore be recovered.

Aluminium chlorohydrate has found wide application in Britain, but the most promising recent developments have been in the field of organic polymers (polyelectrolytes). Certain highly effective and convenient materials are widely used in the USA, but have remained expensive. More recent British products promise marked improvements both in sludge throughput and in cost. Research work in Britain shows that these materials are highly selective in their application. Cost-benefit assessments may be readily made by means of the laboratory techniques already described, which also facilitate day-to-day optimisation of plant operation.

The limited British application of cloth-covered vacuum filters may be attributed to the difficult nature of British sludges, and filter yields of the order of 5 lb/ft^2 h (25 kg/m^2 h) (dry solids) are restricted to one plant with no secondary treatment of sewage, where the ligested sludge is elutriated and also chemically conditioned before iltration. Advances in conditioning techniques may well improve the ituation in the future, but the present design of filters of this type ppears to be insufficiently versatile to permit full exploitation of the otentialities of chemical conditioning.

The solid bowl continuous centrifuge with screw conveyor has in ecent years found increasing application to sewage sludge, particu- arly in the USA, but the removal of suspended solids appears to be imited to about 70 per cent. By prior conditioning with poly- lectrolytes, the recovery may approach 100 per cent, but the mount of conditioner required to overcome the shearing effect of he centrifuge adds considerably to the cost. The increased recovery s also gained at the expense of reduced solids content of the cake. dvances in centrifuge design and in the polyelectrolyte field may, owever, increase the attractiveness of the method in the future.

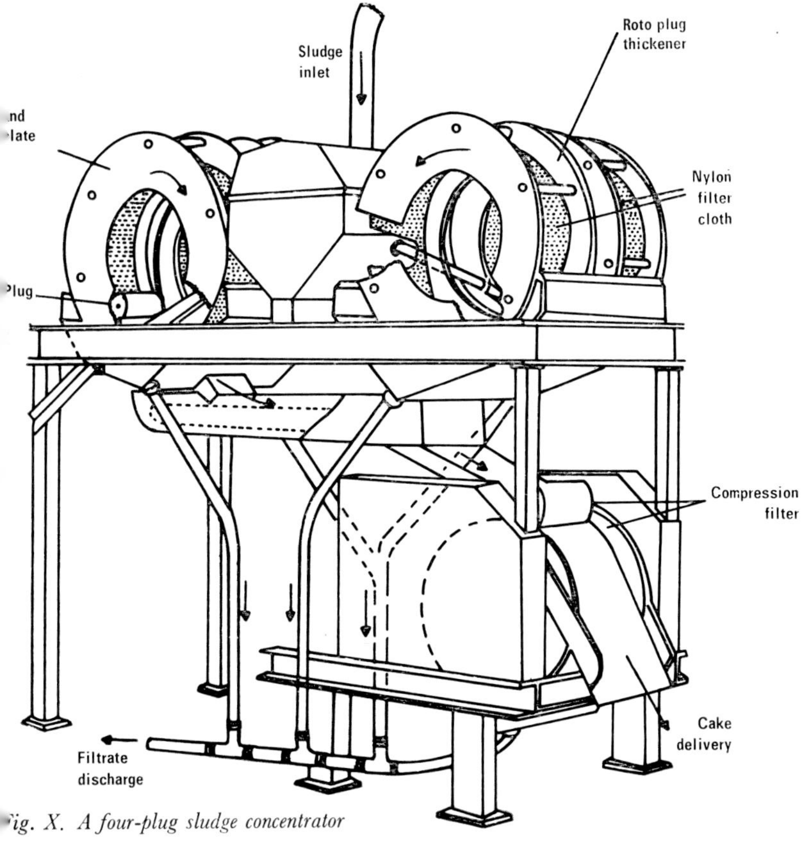

ig. X. A four-plug sludge concentrator

40. A filter press installation

The Roto-plug sludge concentrator, first introduced in the Unitec States, has been further developed by a British manufacturer (Fig. X) Its main application is to crude sewage sludge, but in Britain it ha been used to dewater digested sludge after the addition of paper pulp The recovery of solids is reduced if the sludge contains difficul secondary sludge, although this problem can be overcome by the us of conditioning chemicals.

The oldest British method of mechanical dewatering is filte pressing, in which sludge is filtered after injection, at pressures o about 100 lb/in² (7 kg/cm²), into chambers about 1½ in (40 mm) wid formed between movable plates covered by filter cloths. After filtra tion, the cakes are allowed to drop from the presses by separating th plates. The method has one outstanding advantage: it permits th direct and practically complete removal of suspended solids fron crude sludge, thus avoiding the recirculation of polluting matte inherent in some other methods of treatment. As a result of the highe operating pressure, cakes produced from compressible solids (sewag sludge) have a higher solids content than is achieved by othe methods. Presses are highly versatile in that they can utilise dramati improvements in sludge filtrability, resulting from new conditionin techniques. Labour requirements for separating the plates have bee greatly reduced by automation and mechanical maintenance cost are very low (photograph 40).

t sea

artial disposal by gasification of solids during treatment facilitates he final disposal problem, but even after anaerobic digestion of ludge at large sewage works—for example, those serving London— massive disposal operation remains. The solution adopted at these nd other works is to transport the sludge in ships for dumping at sea. Vhere applicable, the method is economical and final, and wide ritish experience over many years is available for application else-vhere. As an alternative method of transport, pipelines may be used nd feasibility studies have recently been completed for a scheme to erve the London area which, if implemented, would include a ubmerged section extending nearly 8½ miles (13 km) into the outhern North Sea, carrying over 6 mgd (28,000 m^3/d) digested sludge.

n land

or very large communities, where insufficient agricultural land is vailable for economic disposal, or where anaerobic digestion is not easible, a solution may be provided by incineration. Although a on-combustible ash still remains for disposal, it is inert and repre-ents only about 5 per cent of the weight of, for example, filter cake. he two main methods of incineration are in multiple-hearth and uidised-bed furnaces. Both methods appear to permit odourless ombustion with satisfactory recovery of ash from the flue gases. A ew British system of incineration is now available at greatly reduced apital cost. In order to operate without auxiliary fuel, any method epends on efficient preliminary dewatering. It would appear that lter pressing is the most economical, and probably the only method t present available to meet this requirement, combined with near-omplete recovery of solids.

. D. SWANWICK

fter a period in the pharmaceutical industry, Dr Swanwick graduated and arried out postgraduate research at the University of London on the vapour ressure of new alkoxides of certain metals. In 1957 he joined the Water ollution Research Laboratory where, as a principal scientific officer, he leads a am investigating various aspects of sludge technology. His activities range om fundamental studies of the nature and behaviour of sludges to investigation f operational problems. During recent years he has advised on about half the rious problems encountered with the anaerobic digestion process in Britain. Ie has also studied the application of new methods in the USA, and is author f some 20 papers on sludge technology.

CHAPTER 10

British Achievements Abroad

BY JOHN T. CALVERT, CBE, MA, BSc, MS, FICE, MIWE, FRIC, FASCE

Necessity, a proverb says, is the mother of invention. In one importan instance, the father was an Englishman, for it was Sir John Haring ton, a godson of Queen Elizabeth I and a remarkable man in man respects, who designed the first water-closet in 1596. Another Britis inventor, the Yorkshireman Joseph Bramah (1749–1814), perfecte it at the turn of the nineteenth century. As a consequence, Britis engineers were the first to develop the water-carried system o sanitation as we know it today—and the first also to recognise and t solve the problems set by the pollution of rivers and lakes arisin from indiscriminate methods of sewerage.

As the major growth of the British Empire coincided roughly wit those developments, it was only natural that British public healt engineers should have been called upon to exercise their skill in th vast territories forming part of that Empire, with the result tha very many of the larger cities of the world have sewerage system designed and constructed by British engineers. On the continents o Europe and North America, they have not been quite so active, owin to the competence of local engineers in those countries, who had no been slow in availing themselves, as was only natural, of the experi ence and knowledge won by the British pioneers in this field. Neve theless, just before the first world war, British engineers were engage to report on the drainage of St Petersburg, today Leningra Moreover, British manufacturers of sewage treatment equipme have supplied machinery to (among other places) Darmstadt i West Germany, Kitchener and Calgary in Canada, and Brasili the capital of Brazil. In recent months British manufacturers hav supplied equipment for a sewage treatment plant near Paris, an for the enormous Emscher scheme in Germany, where the entir river is to be run through a treatment plant before its waters ente the Rhine.

All over Asia and in large parts of Africa, British engineers hav een responsible for the construction of drainage schemes. One of the irst to be tackled was Bombay, where, by 1860, conditions had ecome so bad that proposals were put forward by the municipal ngineer for the collection of the city's sewage and its discharge into he harbour. Before the turn of the century, schemes were under way or the drainage of Cairo, Calcutta and Rangoon, later followed by rojects for Alexandria, Karachi and Singapore.

British consulting engineers such as Baldwin Latham, Carkeet ames and Shone were concerned with many of these schemes, and it nay be noted that Rangoon was provided in 1889 with a drainage cheme comprising the first installation of Shone ejectors, by which he sewage was pumped into a main collecting channel by the use of ompressed air.

Many of the cities mentioned above are on the sea coast or on large ivers, and the first schemes constituted collection of the sewage and ts discharge into a large body of water where treatment was effected y dilution followed by oxidation by the forces of nature. Cairo, owever, provided an interesting exception to this method, even hough the waters of the Nile were available for dilution of the ewage. There was, at that time, a feeling that the fertiliser value of lomestic sewage should not be wasted, and this led to the establishnent of a sewage farm at Gebel el Asfar, about 15 miles from the

1. Sludge treatment plant in Aire, Geneva

centre of Cairo. At these works the sewage was treated first b sedimentation and then by filter beds before being applied t approximately 3100 acres of sewage farm. The farm has since bee most successful, largely for the growth of citrus fruits and, to som extent, cereals. The growth of many of these cities in recent times ha demanded further treatment of the sewage, as the increasing popula tion has led to nuisance conditions in the watercourses.

After 1945, the upsurge of nationalism in developing and nov independent countries, together with the redistribution of wealt created by the increasing demand for oil, resulted in a rapid growt within these states of modern sanitation facilities and other services High on the list of requirements of these countries are water supply roads, communications and industrial development. The provisio of water calls for enormous and constantly rising expenditure, whic can only partly be met by the World Health Organisation. Th transformation of the British Empire into the Commonwealth ha loosened the ties between Britain and her overseas territories an opened their doors to the operations of engineers from man countries. Nevertheless, the high reputation of British engineers ha won for them a notable share of the work now in progress.

Their work is spread throughout the world, going as far afield a the Solomon and Ellice Islands in the Pacific Ocean, Hong Kon and Brunei in South-East Asia, as well as Jamaica in the Wes Indies. But it is probably on the African continent and in the Middl East that the bulk of their work is located, and they are at presen working on public health schemes in almost every country in tha part of the world. In Libya alone, £24,000,000 worth of work i currently in progress.

In the city of Tripoli, treatment works are being provided for a present population of 120,000, with facilities for extension up to fou times that number and with an ultimate flow of 24 mgd ($1 \cdot 1 \times 10^5 m^3/d$). These works are equipped with the most modern machi nery, including comminutors to break down large solids and detritor for the removal of grit. All the tanks, which are circular, are mecha nically scraped and the sludge removed is digested in heated tanks s as to eliminate nuisance in the works before it is dried on open drying beds. Secondary treatment in these works is provided by rectangula percolating filters with recirculation of the filter effluent, in order t increase the filter loading and thus reduce the cost. An interestin feature here, which has been incorporated on account of the genera water shortage on the North African coast, is a tertiary stage of san

42. Aerial view of the sewage treatment works for the city of Benghazi, Libya, during construction

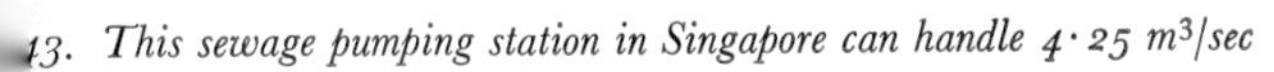

43. This sewage pumping station in Singapore can handle 4·25 m³/sec

44. The Rustamiyah sewage treatment works for the city of Baghdad

filtration, followed by sterilisation through chlorination, in order to make the final effluent safe and suitable for general irrigation.

Another important scheme designed and supervised by British engineers is the complete sewerage and sewage treatment for Baghdad. This city, with a population of over a million, had, until 1960, no sewer whatever; its excellent and copious water supply was disposed of entirely by individual septic tanks overflowing into the ground. New sewers are now in course of construction; those on the east bank of the river Tigris are almost completed; those on the west bank are about to be started. Originally, in view of the adequate dilution available, it was intended to locate the treatment works on the river bank and to provide only primary sedimentation, but, after the 1958 revolution, all suitable riverside land was pre-empted for military purposes. The proposed site of the works had to be moved 10 kilometres up the Diyala river and the effluent discharged into this relatively small tributary of the Tigris, with the result that secondary treatment had to be added to the plans. These works on the east bank of the Tigris are now in full operation and producing an effluent of excellent quality. In addition to scraped circular tanks, heated sludge digestion tanks, and all the other equipment supplied by Britain, they incorporate biological treatment by the diffused air activated-sludge process which operates most efficiently in the warm climate of Iraq. It is now proposed to install a mechanical aeration process on the west bank of the Tigris, so as to obtain a comparison of the economy and reliability of the two methods under Middle East conditions.

South-East Asia is another developing area and Singapore is typical of a city with a growing population and an increasing appreciation of the merits of proper sanitary facilities. Drainage of crude sewage into the rivers had led to most unpleasant conditions, which are now being remedied with the help of British advice. Schemes estimated at about £9,000,000 have been constructed or started since the war. The city's sewage is drained to two treatment plants. One, the Kim Chuan Road Works, serves the eastern part of the city and new development north-east of the Kallang Basin (the Toa Payoh Scheme); the other, the Ulu Pandan Works, serves the western parts of the city. The former is designed to provide eventually for a population of one million and the latter for 600,000 persons. Both works provide full biological treatment by means of the activated-sludge process following grit removal and sedimentation, and the sludge is digested in heated tanks with gas-holder roofs.

The methane evolved in the process is used to generate power for works operation and particularly to drive the compressors for aeration; over 3000 horse-power (2300 kW) is produced at the western works.

A somewhat unusual feature of the Kim Chuan Road works is that the sludge is not dried on the site, but pumped to a separate plant at Serangoon, where final drying takes place. In this way, the part of the sewage treatment process requiring the largest area of land and best isolated from property can be sited well away from urban developments.

The above examples are merely typical of many major schemes in progress throughout the world and are described to emphasise the important part that British engineers are playing in solving the public health engineering problems of the world.

JOHN T. CALVERT

Mr Calvert graduated from Oxford with first-class Honours in chemistry. He was then awarded a Commonwealth Fund Fellowship and went to the Massachusetts Institute of Technology where he studied and graduated in Civil Engineering. During this period in America, he specialised in public health engineering and was able to visit many water supply and sewage treatment works. In 1932 he joined John Taylor and Sons, consulting engineers, as an assistant engineer. In 1944 he was taken into partnership, becoming senior partner in 1966. Mr Calvert has advised the city of Auckland in New Zealand and the governments of Aden, Saudi Arabia and Sabah as well as the World Health Organisation in relation to the drainage of Istanbul.

CHAPTER 11

Research and Information Services

BY A. L. DOWNING, MA, BSc, DSc, CEng, AMIChemE, FInstWPC, FIBiol, FIPHE

A few years ago, expenditure on research and technical investigation into problems of water pollution control in Britain was estimated to be around £1,100,000 a year, of which about 5 per cent was devoted to work in university departments of public health engineering, and the remainder was roughly evenly divided between Government-sponsored research organisations, local authorities and industrial firms. The situation today is similar, though the proportion of effort devoted to work in universities is probably higher.

A major part of Government expenditure is vested in work at the Ministry of Technology's Water Pollution Research Laboratory, which evolved from an organisation founded in 1927 and is now one of the two largest laboratories of its kind in the world. This laboratory, the activities of which are outlined in more detail farther on in this chapter, undertakes research into the effects of pollution on natural waters and into the development of methods for treatment and disposal of the domestic and industrial waste waters that cause pollution. Research on the problems of dealing with effluents from particular industries is also carried out by a number of industrial research associations—bodies financed partly by subscribing member organisations and partly by the Government. Prominent current activities of these bodies include investigations into effluent treatment and water conservation in the industries served by the British Leather Manufacturers' Research Association, by the Paper and Board Research Association, and by a collaborative group of research associations representing various sections of the textile industry. Furthermore, a number of research projects are financed by the Construction Industry Research and Information Association.

Many of the technical departments of larger local authorities, notably for instance the Scientific Branch of the Greater London Council, the Upper Tame Main Drainage Authority, and the Manchester Rivers Department, investigate problems of sewage treatment, especially those of particular interest in the local situation.

A number of the larger British universities, such as London, Edinburgh, Manchester, Birmingham and Newcastle, have Public Health Engineering Research Departments which, besides providing both graduate and undergraduate training (open to British and foreign nationals), carry out important research programmes. Various university departments of chemical engineering, biochemical engineering and applied biology also engage in research relevant to pollution control.

A growing body of more fundamental work, particularly on the long-term subtle biological effects of pollution, is being carried out under the auspices of the Natural Environment Research Council in laboratories for which the Council has a statutory responsibility, such as those of the Freshwater Biological Association and the Nature Conservancy, and through the medium of grants to universities. The Council also exercises an advisory function in respect of work at the Fisheries Research Laboratory (Ministry of Agriculture, Fisheries and Food) at Burnham-on-Crouch, which includes studies of the effects of pollution on shell-fisheries and marine fish, and at the Salmon and Freshwater Fisheries Laboratory of the same Ministry which, among its activities, undertakes research into the toxicity of certain types of effluent to freshwater fish and in the development of standard toxicity tests.

Important co-ordinating functions are also exercised by the Water Resources Board, which advises the Secretary of State for Wales and the Minister of Housing and Local Government, whose principal responsibilities include determination of the measures necessary to ensure that adequate supplies of water can be made available to meet future demands; the Board also initiates research in those areas where more information is needed for the discharge of its responsibilities. Typical of the kind of research activity developing from the Board's initiatives is a comprehensive investigation to establish, by means of an economic model, the best methods of employing the water resources of the catchment of the Trent, one of the largest rivers in the country, to meet the future demands for water in the industrial midlands and in neighbouring areas. Collaborating with the Board in this investigation are staff of the Trent River Authority, the Ministry of Housing and Local Government, the Water Research Association, the Local Government Operational Research Unit, Birmingham University, and the Water Pollution Research Laboratory.

Certain industrial firms manufacturing treatment plant and equipment, and providing effluent treatment advisory services, support these activities with extensive research and development. Among the principal research and development laboratories are those operated by Simon Carves-Monsanto at Cheadle Heath and Ruabon, by Imperial Chemical Industries Ltd at Brixham, and by the Nalfloc-Head Wrightson Effluent Treatment Service, Northwich.

Finally, though this account is primarily concerned with water pollution research it is worth noting that a quite separate large research effort is devoted to the problems involved in water supply by the Water Research Association which commands a budget approaching £300,000.

RESEARCH PROGRAMMES

Every effort is made to ensure that the programmes of the many organisations engaged in water pollution research are co-ordinated in such a way as to avoid duplication. For example, the programme of the Water Pollution Research Laboratory, the largest conducted by any individual organisation in the field, is decided by a Steering Committee which includes the chairman of the Confederation of British Industry's Water and Effluents Committee, a senior member of the Ministry of Housing and Local Government, the director of the Water Resources Board, and the secretary of the Natural Environment Research Council.

This Steering Committee is advised by six committees, including representatives of the various bodies with an interest in the Laboratory's work. Three of these are joint committees with the Confederation of British Industry, with the Associations of River Authorities and Scottish River Purification Boards, and with the Institute of Water Pollution Control, this last representing principally the interests of local authorities. In addition, there is a new Advisory Sub-Committee on Treatment Plant which includes representatives of plant manufacturers and of consulting engineers, a Basic Research Committee, which advises on fundamental work and helps to maintain contact with relevant research in universities, and a Coastal Pollution Research Committee.

Space will not permit an account of the research activities of all the bodies that have been mentioned. Attention is confined to a few of those with large programmes.

The Water Pollution Research Laboratory

The main objects of the Laboratory's current research work are to provide improved methods for the scientific management of natural waters (especially rivers, estuaries and coastal waters), based on a detailed knowledge of the effects of pollution on water quality, and to develop and improve methods of treating sewage and all types of industrial waste waters. A broad range of expertise is required in this research, and some 110 technical staff, trained in various branches of science and engineering, are employed, many of them in multi-disciplinary teams.

The Laboratory is organised in three divisions; two are concerned directly with the main lines of research already mentioned; the third, a service division, provides central services and carries out extensive development of new instruments, analytical methods, radiochemical techniques, and computer programs for data processing and mathematical simulation of the wide range of phenomena under study.

One of the main projects in the work of the Pollution Division, accounting for about one-sixth of the total expenditure of the Laboratory, is concerned with the development of methods for determining the degree of pollution which would result from the discharge of effluents into coastal waters at any given point, thus allowing an accurate assessment to be made of the most effective and economical method of disposal of waste waters and sludges from any given coastal area. The work involves measurement of the dispersion of effluents in the sea, using radioactive isotopes and readily identifiable bacteria and bacterial spores as tracers, study of the factors governing the survival of bacteria in the sea and development of mathematical models that can be used to predict the degree of pollution from particulars of the discharge and the hydrography of the area (photograph 45).

An important continuing line of research is concerned with the development of methods for assessing the effects of pollution on the distribution of dissolved oxygen in estuaries. The type of mathematical simulation used by the Laboratory for predicting the effects of polluting discharges on the condition of water in the Thames estuary, to which reference has been made in earlier chapters, is being extended with some success to other estuaries, including that of the Tees, in which the situation is more complex than in the Thames because the water is vertically stratified. So far as is known, this is the first occasion that a predictive model has been developed for this type of estuary.

45. Study of the rate of dispersion of sewage in coastal waters using scintillation counters in submerged towing vessels to measure the activity of radiotracers discharged from the outfall

6. Recirculating channel for studying factors governing the rate of self-purification of rivers

Similar methods are being developed to determine the distribution of dissolved oxygen in polluted rivers (photograph 46), much of the present effort being concentrated on establishing conditions which give rise to the formation of slimes and to deposition of suspended matter to form mud deposits, since experience has shown that in the slow-moving rivers often found in Britain, these factors frequently dominate the oxygen balance.

Another important aspect is the determination of the effects of the oxygen resources of natural waters of the photosynthesis and respiration of rooted plants and of planktonic algae. As in several other countries, some natural waters in Britain are becoming increasingly enriched with plant nutrients such as nitrogen and phosphorus (a process known as eutrophication), mainly as the result of discharge into them of effluents and of the run-off from agricultural land. In some countries, the process has given rise to excessive growths of algae, which can cause various problems, including difficulties in the treatment of water for public supply, and unsightly rotting debris when conditions become unfavourable for growth; moreover, certain blue-green species release poisonous toxins which have been responsible for mortalities among fish and livestock. So far, difficulties in Britain have been fairly few, but the implications of current trend need further examination. An increasing effort is therefore being devoted to the determination of the main sources of nutrients in rivers and of the effects of these nutrients on algal productivity.

Both in rivers and estuaries, fisheries are important. To assess how conditions can best be kept suitable for fish in water subject to pollution, investigations are being made of the toxicity of individual poisons and mixtures of poisons both in the laboratory and in the field. Considerable progress has been made in the interpretation of the toxicity of polluted natural waters from measurements of the concentration of the poisons that they contain. Much of the past work has been concerned with the more immediately acute effects of poisons to trout, but increasing attention is now being directed to effects on coarse fish and the more subtle long-term effects and interactions of constituents of effluents.

Work on the treatment of waste waters and the sludges derived from them accounts for about half the total effort of the Laboratory. It is concerned with the elucidation of the basic mechanisms of the commonly used unit processes, including the fundamental microbiology of biological processes, so as to provide a sound basis for optimising the design of treatment plants; also with the remedying of operating difficulties, especially at sewage works, with the development of new processes as need arises, and with the improvement of monitoring and control systems.

A considerable proportion of this work is devoted to the study of the biological oxidation processes employed in the treatment both of sewage and of organic industrial effluents and to the treatment and dewatering of domestic and industrial sludges by both traditional and more modern methods (photographs 47/8). Much progress has been made in unravelling the kinetics of microbial growth and oxygen transfer in aerobic systems and in the development of high-rate treatment processes. Recent work on sludge technology has been particularly encouraging and has led already to substantial improvements in performance of certain types of mechanical dewatering equipment.

The growing use of new chemicals, both in industry and in the home, has increased the incidence of difficulties arising at sewage works as the result of the discharge of such materials to drainage systems. Among these difficulties have been the now well-known problems resulting from the presence of synthetic detergents in

47. Blocks of pilot-scale percolating filters used for studying the effects of operating and design variables on filter performance

48. Small-scale activated-sludge plant for treatability studies

sewage and the inhibition of aerobic and anaerobic processes by various other substances. Many of these problems have been solved by work in which the Laboratory has played a leading part, and although further problems will doubtless arise in future, the techniques that have been developed, for example for diagnosing inhibition and for detecting the source of inhibitors, and for establishing biodegradability, should allow many difficulties to be dealt with more swiftly than was possible in the past.

Because of increasing pressure of demand for additional water supplies and for the greater exploitation of rivers subject to pollution for recreation, much attention is being given to the development of methods of tertiary treatment for producing effluent of higher quality than is obtained from conventional biological treatment plants, and especially to the development of water-reclamation processes for the production of a final treated effluent suitable for re-use (photograph 49).

The Laboratory has already demonstrated that water suitable for many industrial purposes can be produced from sewage effluent at a reasonable cost in a number of ways. An increasing effort is now being devoted to the possibilities of developing economical methods of recovering water of even higher quality, possibly even suitable for

49. Pilot plant for study of the reclamation of water suitable for industrial use from sewage effluent

potable use. One aspect of this work being investigated in collaboration with the Process Technology Division of the United Kingdom Atomic Energy Authority, Harwell, involves the study of potential applications of the relatively novel process of reverse osmosis, which entails forcing the liquid under pressure through a thin membrane which retains impurities.

Sponsored investigations

In addition to the approved programme of research, the Laboratory also undertakes work within its field for individual sponsors on payment of a fee, this service being available to organisations overseas as well as to those in Britain. Considerable use is made of these facilities especially by industry, local authorities and their consultants, particularly for investigations into problems that can be solved relatively swiftly by well-established techniques. A frequent requirement is for investigation of the treatability of particular sewages or industrial waste waters to provide a basis for plant design, and the Laboratory maintains a range of laboratory-scale and pilot plants for such studies, including mobile units which can be installed on the sponsor's premises if required.

Information services

As much as possible of the Laboratory's work is published, the majority in scientific and technical journals, the total of papers by members of the staff now being close to 600. Reprints of most of those published within the last 10 years can be obtained from the Laboratory on request. Also available on request are copies of the Laboratory's quarterly *Notes on Water Pollution*, which are short summaries of the state of knowledge on particular aspects of this problem. In addition, the Laboratory produces Water Pollution Abstracts, Annual Reports, and technical papers (these last describing investigations too large to be fully reported in journals), all of which are sold by Her Majesty's Stationery Office, London, and can be obtained through various agencies abroad.

An Information Service on Toxicity and Biodegradability (INSTAB) enables an inquirer to obtain a considered account of the available information on the behaviour of individual substances in rivers and biological treatment plants. A list of substances on which information is available, more than 1500 in number, is obtainable from the Laboratory on request. A charge is made for inquiries from abroad and for those in Britain requiring more than a day of a man's time.

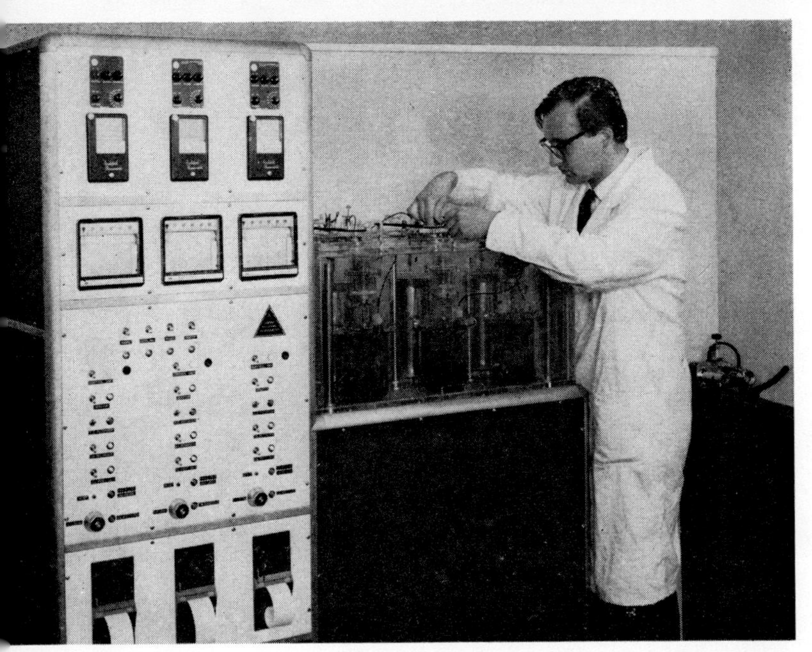

50. Simcar recording respirometer for study of the treatability and biodegradability of effluents

Biochemical Engineering Division of Simon-Hartley Ltd

The work carried out has been directed over the last 10 years towards the practical problem of predicting the design and performance of effluent and sewage treatment plants to suit specific cases, to understand the treatment processes sufficiently to permit a guarantee of performance to be given, to cure difficulties arising during operation and to improve existing processes and equipment.

One of the work's more important aspects has been the development of an automatic recording respirometer for use on activated-sludge systems (photograph 50). A close liaison with Monsanto Chemicals Ltd, who were developing similar equipment at the same time, has resulted in a piece of equipment capable of producing accurate results with the minimum of attention, and subsequently to confirmation that, under plant operating conditions, the oxidation reaction in an activated-sludge system is essentially zero order.

Other work has included the development of a simple test for the detection of toxic metals in effluents or activated sludge, and various other 'trouble-shooting' techniques.

51/2. Experimental pilot-scale high-rate composite biological filters at the Water Pollutio Research Laboratory. These are used to study the performance of different types of packings

53. The Brixham Research Laboratory's pilot-scale plant at Buckfastleigh for developmen trials of plastics filter packings

Brixham Research Laboratory

This laboratory (now an integral part of ICI's Engineering Services Department) has for many years offered an investigation and advice service to all manufacturing divisions of the company on effluent treatment and disposal and on the prevention of pollution by industrial wastes of rivers, estuaries and coastal waters.

At present, the programme includes research on estuarine and off-shore pollution, on treatment of industrial effluents and domestic sewage (particularly by biological filtration using plastics media) (photograph 53), on solid-liquid separations, including the application of polyelectrolytes in sludge dewatering, on the toxicity of wastes and products to fish and other aquatic organisms and on their biodegradability, on the toxicity of industrial effluents to aerobic and anaerobic biological treatment processes, and on the application of respirometry.

A. L. DOWNING

Dr Downing joined the Water Pollution Research Laboratory in 1946 after graduating in Natural Sciences at Cambridge. In 1947 he was seconded to the Fisheries Research Laboratory, Lowestoft, for a year and, following this, spent 18 months at the Government Chemist's Laboratory, London. While there he obtained a BSc special degree in chemistry from Birkbeck College. He rejoined the Water Pollution Research Laboratory as a scientific officer in 1950 and worked for about three years on the fate of radioisotopes during treatment of domestic water supplies. Subsequently, he was engaged in studying the dynamics of aeration processes and the kinetics of biological oxidation and nitrification in treatment plants and in natural waters. He was appointed deputy director of the Laboratory in 1964 and director in 1966. In the following year he was awarded the degree of DSc by London University for his work in biochemical engineering. He is author of some 50 technical papers.

CHAPTER 12

The Role of the Consulting Engineer

BY J. D. WALL, BSc, FICE, FIPHE, MConsE

Part of the purpose of this book will have been served if it has give
some information to those responsible for the health and well-bein
of people as to the means available for the treatment and disposal o
organic wastes and the control of water pollution. The problem
involved and their solution are so diverse and complex that the reade
may well find himself baffled in his attempts to evaluate the particular needs of the community whose health may be his own vita
concern.

Those faced with problems of that nature may be nationa
governments—particularly in developing countries; municipa
authorities—whether they be in charge of large cities or of sma
villages; the controlling bodies of hospitals, schools and simila
institutions; manufacturers whose factories produce harmful effluents; and many others. A few of these, such as the largest municipa
authorities and manufacturing concerns, may be aware of thes
problems and capable of solving them. The majority, however, wi
require guidance at all stages, from the basic conception of a scheme
its planning and construction, the eventual commissioning of th
works involved and—not least—their cost.

Guidance of that kind is the main function of the consultin
engineer. He does not claim to be the only source of such advic
Nevertheless, it is true to say that the majority of schemes for th
control of water pollution are conceived by consulting engineers an
carried out under their control. Many of them are British firms, an
it is relevant to note that, early in 1969, members of the Associatio
of Consulting Engineers were responsible for about £400,000,000 c
work in this and related fields throughout the world.

Why consulting engineers are best qualified to offer advice an
assistance in work of that nature will be briefly outlined in th
present chapter.

Experience

Consulting engineers are, by virtue of long practice and past experience, experts in the appreciation of the nature of a problem, and can quickly determine, in general terms, its probable solution. Following that basic appraisal, they can apply themselves and their skilled assistants to the details of a scheme, again drawing upon their experience of numerous previous problems of a similar nature, as well as upon their study of the science of engineering and of the achievements of the engineering profession.

Knowledge

While the consulting engineer relies mainly on his own knowledge, there are nevertheless occasions when he must enlist the services of specialists with the particular information and equipment required to carry out certain investigations and tests. He knows from experience what special services are required and where they may best be obtained. The following list indicates the type of special investigation that may be necessary.

Land surveys. Particularly in the case of underdeveloped countries, surveys by aerial photography may be required.

Foundations. Where heavy foundations are necessary, the nature of the underlying strata must be ascertained, for which purpose specialist boring contractors may be employed to obtain samples at various depths.

Soil mechanics. The strength of the soils on which structures are to be founded, or from which they are to be formed, may have to be determined by laboratory tests.

Chemical analyses. Certain soils contain substances that may necessitate special forms of construction. The nature of the ground must therefore be ascertained before final design can proceed. Similarly—where sampling is possible—the nature of any liquid to be conveyed in a sewerage system must be ascertained by analysis, not only for the protection of the structure which will contain the liquids, but also to enable the method of biological treatment to be matched to the nature of the particular liquids involved.

The greater part of the work involved in schemes of the kind discussed in the foregoing pages is likely to be civil engineering. It is possible, therefore, that overall control will be entrusted to a consulting civil engineer. Such schemes, however, generally incorporate pumping and other machinery. Most consulting firms of importance have on their staff qualified mechanical and electrical engineers, capable of designing and supervising the construction of this part of the project. When this is not the case, or where the mechanical and electrical part of a project forms a large part of the whole, or is particularly complex, the consulting civil engineer may advise his client to engage an electrical and/or mechanical consulting engineer to act in association with his own firm.

Control of works under construction

The consulting engineer's duties do not cease when he has completed the design of the works and has produced the detailed drawings for their construction. He must then guide his client in the appointment of a contractor to build them, as well as in the selection of plant and equipment. He is responsible for the supervision of the construction of the works, and for the engagement and direction of the engineering staff who represent the client's interest on site. He must control and certify the accuracy of the client's payments to the contractor. Finally, he must test the completed works and hand them over to his client in proper working order. All these duties are within the day-to-day experience and competence of the consulting engineer. Contractors and manufacturers are accustomed to carrying out contracts under this form of control: usually, mutual respect and confidence exist between them and consulting engineers. The latter are nevertheless able, by virtue of experience, to act in the best interest of their client when differences or disputes arise.

Procedure

A potential client will generally know the nature of his problem. Where large communities are concerned, it is frequently the municipal engineer and surveyor who has reported to his authority, and may himself be competent to design a scheme and to carry it out. More often, however, an authority has insufficient staff to take on that extra burden and decides to appoint consulting engineers. Thereafter, the procedure will vary according to the nature of the problem, but the usual pattern is as follows:

(1) The consulting engineer makes the necessary local investigations and preliminary surveys and reports to his client on the need for a scheme, the form of the scheme best suited to meet the need, and the cost.

(2) The consulting engineer's report may include recommendations for the execution of certain works. If the client decides to accept them, he appoints the consulting engineer to carry out the design and supervise the construction of the works. The consulting engineer then makes a more detailed survey of the site of the proposed works, prepares working drawings and contract documents to enable tenders for the works to be obtained, and advises the client as to the choice of contractor.

(3) Work on site starts, supervised by the consulting engineer and also by a representative resident on site, with, on larger projects, the necessary assistants. This resident staff is responsible to, and controlled by, the consulting engineer, who himself pays frequent visits of inspection.

(4) At intervals throughout the work and at its completion, the consulting engineer examines the cost of the work as submitted by the contractor, and certifies it for payment by the client.

(5) Finally, the consulting engineer tests the whole of the works, satisfies himself that they are in good working order, and hands them over to the client.

The foregoing summary is of course no more than an indication of the way in which the whole course of a project may be directed by the consulting engineer acting as his client's professional adviser.

Overseas clients

While all that has been said above applies equally to projects carried out on behalf of clients overseas, there are many other ways in which their special requirements can be met by the consulting engineer.

It is frequently desirable to ensure the rapid progress of overseas work by placing orders for equipment, machinery, manufactured goods, etc., in advance, so that they may be available on site at the start of the work. The consulting engineer will prepare specifications, ask for quotations and place orders on his client's behalf.

Supervision of work overseas frequently requires the employment of expatriate resident engineers. The consulting engineer is usually able to secure the services of men who have previously worked under him and for whose ability he can vouch. He can also undertake the administration of expatriate staff in matters such as housing, travel, home leave, etc.

In the case of developing countries particularly, the need to train their own nationals in the methods of executing work and in the operation of completed plants is often met by arranging with the consulting engineer for the training of suitable men, even to the extent of a period of training in the consulting engineer's head office. He can also arrange for practical experience in operating techniques at sewage works in Britain.

When major schemes of sewerage and sewage treatment have been carried out in a developing country, it is generally necessary to formulate by-laws to regulate the use of the facilities provided by the scheme. Because of his experience in such matters, the consulting engineer can help his client in what may be to him an unfamiliar exercise.

In these and other ways, the consulting engineer will wish to assist so that his client may derive the maximum benefit from the work he has undertaken.

Remuneration

Consulting engineers are paid for their services in the form of fees agreed by negotiation with their clients. The basis of remuneration is laid down by the Association of Consulting Engineers, but is subject to variations according to special circumstances—particularly in the case of overseas clients.

In establishing a basis of remuneration, the Association seeks to ensure that consulting engineers are appointed according to their suitability for the work involved, and not on considerations of financial competition between its members.

Independence

The advice of consulting engineers is designed to ensure that the best materials and equipment available for efficient operation and economy are used in the works under their control. Their loyalty is to their clients alone, with whose satisfaction their own best interests are indissolubly bound up.

They are forbidden by the rules of their Association to participate in any way in the commercial activities of any contractor engaged to carry out the works, or of any supplier of materials and equipment to be incorporated in the works. Their judgment regarding matters affecting the works is thus directed solely to the client's interests.

Responsibility

The value of the services rendered by a consulting engineer rests upon his ability to assume responsibility for all phases of the work he has undertaken on his client's behalf. From his earliest years in the profession, his whole training and experience is conditioned by his success in accepting an ever-growing measure of responsibility. Indeed, his very position as a partner in a firm of consulting engineers is achieved only because of that all-important qualification, which is the factor most highly valued by the clients who entrust him with their commissions.

Information regarding consulting engineers

The Association of Consulting Engineers, with offices at 2 Victoria Street, London SW1, assists potential clients in all matters concerning the engagement of consulting engineers. It defines in detail the conditions of engagement, scales of fees and model forms of agreement between client and consulting engineer, with particular reference to works in Britain. It recognises, however, that for works overseas conditions vary from country to country, and that terms of engagement must therefore be negotiated to suit the particular circumstances.

J. D. WALL

Mr Wall graduated from Manchester University with a BSc in civil engineering and in 1922 joined the firm of Howard Humphreys & Sons, consulting civil engineers, as a pupil of its founder. He then became engineering assistant and held this position until he was taken into partnership in 1945. As a consultant, he has advised on problems connected with water supply, drainage and sewerage, road and bridge construction and general civil engineering. Mr Wall, now retired, has had considerable experience of works overseas.

The photographs in this book are reproduced by kind permission of the following individuals and organisations: The Director of Public Health Engineering, Greater London Council (frontispiece); Water Pollution Research Laboratory (photographs 2, 3, 17, 18, 19, 38, 45, 46, 47, 48, 49, 51, 52); J. D. & D. M. Watson (4, 5, 6, 7, 8, 9, 43); Rock & Taylor Ltd (10); Activated Sludge Ltd (24, 25); Ames Crosta Mills & Co Ltd (12, 14, 27, 28); The Engineer and Surveyor, Farnborough Urban District Council (15); Imperial Chemical Industries Ltd (20, 21, 53); The Permutit Co Ltd (22, 34); G. D. Peters & Co (Engineering) Ltd (23); William E. Farrer Ltd (26, 36); Simon-Hartley Ltd (29, 30, 50); Ward, Ashcroft & Parkman (31); Parsons, Brown & Partners (32, 33); Ferodo Ltd (35); Norstel & Templewood Hawksley Ltd and the Burgh Engineer, Galashiels (39); Johnson-Progress Ltd (40); Norstel & Templewood Hawksley Ltd (41); Howard Humphreys & Sons (42); John Taylor & Sons (44); Lea Recorder Co Ltd (11).

The diagrams in chapters 8 and 9 are reproduced by courtesy of William Boby & Co Ltd (Fig. VI), Glenfield & Kennedy Ltd (Fig. VII), Ames Crosta Mills & Co Ltd (Fig. VIII), Activated Sludge Ltd (Fig. IX), and Davey, Paxman & Co Ltd (Fig. X).

Appendix 1

The following list includes most of the important British manufacturers of plant and equipment for water pollution control and sludge dewatering.

ACTIVATED SLUDGE LTD
41 BUCKINGHAM PALACE ROAD — *telephone: 834 5973*
LONDON SW1, ENGLAND — *telegrams: Acsludgeti London SW1*

Activated Sludge Ltd, the original sewage machinery contractors for the process, are suppliers of specialist machinery developed to equip aeration tanks with the most efficient form of aeration consistent with overall economy.

The seven-inch (18 cm) diameter porous Alundum dome diffuser has been in use for some years and is the result of decades of experience and experimental work which has continued since the discovery of the process in 1914.

Recent development work has led to a more efficient disposition of the diffusers in aeration tanks, known as the high density or multi-line arrangement, giving an efficient and gentle aeration to the tank contents by disposing the diffusers over the whole floor of the tank.

This arrangement not only effects a high oxygenation efficiency (13·5 per cent), but also provides a low sludge volume index (55–75), allowing the mixed liquor suspended solids (active mass) to be maintained up to values of 7000 milligrams per litre commensurate with overall process efficiency. As a result, fully nitrified effluents of high quality are consistently achieved, surplus activated-sludge volumes are of a low order and wastes of difficult treatment characteristics can be more easily handled.

The diffuser can be used for all forms of activated-sludge plants from high rate to extended aeration with consequent benefits at all stages of the process.

Activated Sludge Ltd frequently act as machinery contractors for all aeration plant equipment and can supply air blowers, air filters, electrical machinery, control and measurement systems, together with all necessary liquid control penstocks, sluice valves and handstops as necessary.

As the firm is a subsidiary of Norstel & Templewood Hawksley (members of the Hawker Siddeley Group), specialist machinery for a complete municipal sewage treatment plant can be offered from a single organisation with an international reputation.

AMES CROSTA MILLS & CO LTD
HEYWOOD, LANCASHIRE
ENGLAND

telephone: Heywood 69091
telegrams: Purify Heywood
telex: 63410

Ames Crosta Mills & Co Ltd, a member of the Woodall-Duckham Group, design and manufacture the Simplex range of mechanical equipment which covers all stages of the sewage purification process from screening and grit extraction to sludge treatment. They are able to supply either individual items or complete plants of all sizes, from small community installations to those serving the needs of the largest cities. The firm has also specialised for many years in the provision of effluent treatment plants for industry.

Ames Crosta Mills are represented by subsidiary companies, licensees or agents in more than 100 countries and, over a long period of producing equipment for overseas markets, have acquired a fund of experience which ensures that Simplex designs suit the climatic conditions peculiar to each locality.

In particular, Ames Crosta Mills are known throughout the world for their Simplex High Intensity Aeration Process which has proved to be both efficient and reliable. There are over 1000 Simplex Aeration Plants in use in locations varying from the high and dry conditions of Iraq to the sub-zero temperatures of Canada. They include installations in:

	Design Dry Weather Flow
London, England	100 million Imp. gallons daily (454,500 m^3)
New Delhi, India	30 million Imp. gallons daily (136,300 m^3)
Kitchener, Canada	13·5 million Imp. gallons daily (61,360 m^3)
Brasilia, Brazil	10 million Imp. gallons daily (45,450 m^3)
Brampton, Canada	3 million Imp. gallons daily (13,630 m^3)

Mention may also be made of two other Simplex Plants: one, for Calgary, Canada, with a dry weather flow of 72 million gallons daily (327,000 m^3) will be the largest mechanical aeration plant so far completed in North America; the other, the largest in Europe, will treat the entire flow of the river Emscher in West Germany, which at peak may be 500 million gallons a day (2,270,000 m^3).

The process is unique in being suitable for very deep aeration tanks where land is at a premium, since the Simplex draught tube ensures positive mixing irrespective of tank depth. For maximum flexibility in plant design, a full range of aerating cones is available, powered by individual drive units of from 2 hp (1·5 kW) to over 100 hp (75 kW). A mobile Simplex floating cone is supplied for applications outside the sewage works, e.g. in lagoons, lakes and waterways, where a conventional aeration plant would be impractical.

DORR-OLIVER CO LTD
NORFOLK HOUSE
WELLESLEY ROAD
CROYDON CR9 2DS
ENGLAND

telephone: 686 2488
telegrams: Dorroliver Croydon
telex: 27118

The company designs and installs a complete range of systems and equipment for the control of water pollution. The specialised range of Dorr-Oliver equipment includes:

The Dorrco screen and disintegrator combination, designed to remove and macerate all coarse screenings from the sewage flow and return them to process without manual handling of any kind. The screen is available in standard sizes from two feet (61 cm) wide by three feet (91 cm) deep to eight feet (2·4 m) wide by eight feet (2·4 m) deep. Special designs for larger units are available.

The Dorrco fine screen consists of a drum with perforated plates and is designed to remove screenings from sewage, trade wastes or river water prior to subsequent treatment. It has a particular application on sea outfalls and the use of this unit enables solids to be removed from the flow for separate disposal. Dorrco fine screens are available in sizes from three feet (91 cm) diameter by three feet (91 cm) face to 16 feet (4·9 m) diameter by 9 feet (2·7 m) face.

The Dorr detritor is designed to remove and wash the grit and sand normally present in sewage flows and discharge it in a clean and relatively dry condition for use or disposal as required. The machine consists essentially of a grit-collecting mechanism together with a cleaning and discharging mechanism and is available in standard sizes for all flow ranges.

Dorr clarifier units are available with either a centre-driven rotating scraping mechanism, or a rotating half bridge, designed to move solids settling in a circular concrete tank towards the centre, from where they are discharged either by gravity or pumping in the form of thickened sludge. Standard units are available up to 200 feet (61 m) diameter. Special designs can be made for larger units.

Dorrco distributors are available in many different combinations and are designed for long and efficient service under a large variety of conditions, self-propelling, with either two or four arms, particularly applicable to large diameter percolating filters.

Dorr-Oliver aeration systems are available for activated-sludge treatment for either coarse bubble or mechanical aeration with specialist design and laboratory services available.

Dorr-Oliver special purpose pumps include suction pumps for removing, by positive displacement, thickened underflows from sedimentation and thickener tanks, diaphragm slurry pumps and the Dorrco plunger pump, specially designed for handling sludges in sewage and industrial waste treatment plants.

Dorr primary digesters are available for anaerobic digestion of sewage sludges in concrete tanks with gas-holder or fixed concrete roof designs. Complete mixing is obtained by integral circulatory pumps, and sludge heating is carried out by water to sludge heat exchangers, either within the tank or in the heater house; the necessary heat requirements are provided by methane gas-fired boilers. Sludge scraping mechanisms can be provided for the primary stage to suit flat floor digesters.

Dorr secondary digesters, also with rotating sludge scraping mechanisms, are available for installation in concrete tanks with a relatively flat floor, where a secondary stage is required in sludge digestion plants.

Vacuum filters—Dorr-Oliver offers a complete range of rotary vacuum filters, designed to dewater sludge from sewage and industrial waste treatment plants effectively, with various methods of cake discharge.

Centrifuges of which various types are available for dewatering sewage and industrial waste sludges.

Dorr-Oliver FS disposal system is an efficient, controlled process for the complete disposal of organic sludges and combustible materials. The system has the advantage of very small area requirements, low maintenance and operating costs, no moving parts inside the incineration unit, strict odour control and no air pollution.

Phosphate extraction process, a system for extracting phosphorus from sewage and trade waste effluents, to prevent the growth of algae in the receiving waters.

Dorr-Oliver RapiPress system for high-pressure, short-cycle filtration—a new, fully automatic system for the separation of the solid and liquid phases of a slurry, available in two forms: as a slurry unit for fluid pumpable feeds, and as a semi-solid unit for feed material, which will remain on a flat plate with a reasonable angle of repose.

WILLIAM E. FARRER LTD
CROWN WORKS, WELBY ROAD
HALL GREEN, BIRMINGHAM 28
ENGLAND

telephone: 777 3381-8
telegrams: Farrer Birmingham 28

William E. Farrer Ltd has specialised in the design and manufacture of sewage purification equipment for over 65 years. During that period the firm has built a comprehensive range of equipment to meet almost every aspect of sewage treatment, from the smallest installation up to the sophisticated and highly mechanised equipment for modern sewage treatment on large works.

From its early days in the sewage field, the company has brought a progressive and forward-looking outlook to problems associated with sewage treatment and has been responsible for pioneering many developments now accepted as commonplace in modern design.

Included among its products is a complete range of cast-iron penstocks, ranging from the very smallest valves of three-inch (75 mm) diameter to penstocks in use on power stations up to 10 feet square (3 m square). Many of these can be supplied with exclusive features of adjustments, which time and site experience has proved satisfactory.

A very comprehensive range of equipment is available for biological filters, including self-dosing types for small isolated works, up to power-driven distributors for the largest schemes. These distributors include many patented features which have given satisfactory performance and trouble-free operation for many years.

The firm's range includes travelling distributors for rectangular filter beds, where simplicity of operation and design, as well as controlled periodicity of distribution, are important features.

For over 30 years, the company has pioneered the use of mechanical de-sludging equipment for sedimentation tanks, both circular and rectangular, installing the first rectangular Mieder type scrapers at the Mogden sewage works, which was the first installation of mechanical scrapers adapted to rectangular tanks in Britain. Scraper equipment can now be supplied in steel and aluminium fabrication to cover any application for which this equipment is required.

In recent years the company has developed and supplied to many large works the patent Buoyant Flight type of scraper, which provides the simplest means of de-sludging and de-scumming rectangular settling tanks, avoiding the use of complicated electrical equipment by reason of its non-reversing features. These units can be maintained with a minimum of attention; no unsightly structures are visible above the top of the tanks. Installations in continuous service over the last 12 years show negligible chain wear.

Several works are now provided with the Farrer Jet Aeration process of biological sewage purification, which utilises low air pressure, permitting blowers instead of compressors to be used in conjunction with a unique design of jet which is completely unchokeable. This is a 'coarse bubble' system, suitable for both sewage and industrial waste treatment. Capital costs are low and installations have proved capable of producing well-nitrified effluents in short retention periods.

In the field of sludge treatment, the company took over in 1953 the original patents of the Porteous Company and has since developed the original heat treatment system into a process of continuous operation, which provides the most positive means of treatment prior to sludge dewatering available today. Many installations now in process of construction are based on the successful design operating at the Morley sewage works. Chemical additives are unnecessary, and any type of sewage sludge or admixture of sewage sludges can be treated with complete success, producing a dry, sterile cake as an end product. Sewage sludges made amenable to dewatering by the heat treatment process are in the optimum condition for incineration where this is required.

HAM, BAKER & CO LTD
CLAY LANE, OLDBURY
WARLEY, WORCESTERSHIRE
ENGLAND

telephone: 552 5901-7
telegrams: Penstock Warley

Ham, Baker & Co have been providing equipment for water control and sewerage schemes for over 80 years and now specialise in the manufacture of valves, penstocks and ancillary items. The engineering works have also iron and non-ferrous metal foundries capable of producing castings up to 10 tons (10,160 kg), thus enabling manufacture of complete valves and penstocks up to approximately 10 feet square (3 m square). Above this size, steel fabricated structures can be supplied.

All Ham, Baker & Co products undergo thorough tests and all valves are hydraulically tested before despatch.

The range of valves and penstocks is as follows:

Valves: air release, butterfly, disc, flap, float, foot, hydrant, reflux, sludge, wedge gate.

Penstocks: cast iron, aluminium, disc flushing, hand lifting.

Other products are: air and pressure covers, bell mouths, handstops, hydrants (pillar), pipe fittings, screens, sewage hatchboxes, spigot and socket adaptors, subsoil relief valves, surface and meter boxes, tidal flaps, timber gates, ventilators, weirs and notches.

NORSTEL & TEMPLEWOOD HAWKSLEY LTD
2 BUCKINGHAM AVENUE
SLOUGH, BUCKINGHAMSHIRE
ENGLAND

telephone: Slough 26911
telegrams: Temphawk Slough
telex: 84251

The firm, which is a Hawker Siddeley company, specialises in the design and manufacture of a wide range of plant for the sewage and water treatment industries. Tank scraping machines to suit all requirements are made at the company's Slough factory and, in particular, light alloy designs to reduce weight and maintenance painting have been pioneered and brought to the point of general market acceptance.

The firm was granted a Queen's Award for Technical Innovation in 1968 for its work on the development of sludge lifting machines. These machines have mechanised the distribution of digested sludge on to drying beds and its subsequent removal by a cutter lifter supported from the bed walls without interference to the under-drainage media. Adaptations are available for beds of from 20–60 feet (6–18 metres) span including old beds of pre-mechanisation design, provided these are of regular plan form. Major installations are operating at the West Herts, Rye Meads and three Greater London Council works and also at Paris where a second large machine is to be installed shortly. Similar machines have been developed for the re-sanding of slow sand filters for the removal of impurities at water works.

The company has devoted much attention to the sludge dewatering problem, and has developed the continuous Porteous process of conditioning by heat to provide a compact physical and mechanical system not subject to weather hazards or requiring large land areas. Plants of this type are now operating in Switzerland and Britain and others are under construction in the USA, Germany and South Africa. Incineration plant for screenings and dried sludge is also available, the latter being under licence from BSP Corporation of San Francisco for British and some overseas markets.

Two specialist subsidiaries add to the company's range, namely Activated Sludge Ltd, pioneers and leaders for 50 years of the diffused air process for biological treatment, and F. W. Brackett and Co, designers and manufacturers of screens and strainers for the water and sewage industries. From these facilities comprehensive treatment plant schemes can be offered and new developments are constantly under review and preparation.

Dual-fuel engines to run on oil or methane are another Hawker Siddeley product, and many of these are installed on sewage works to make use of the gas produced from sludge digestion and to convert it to electric power. The engines are made by Mirrlees National of Stockport and Lister Blackstone of Stamford.

SIMON-HARTLEY LTD
STOKE-ON-TRENT ST4 7BH
ENGLAND

telephone: Stoke-on-Trent 24361
telegrams: Simhart Stoke-on-Trent
telex: 36305

Simon-Hartley is a relatively new company formed by the merging of Hartleys (Stoke-on-Trent) Ltd and the Biochemical Engineering Division of Simon-Carves Ltd. In it are combined the long-established experience of the former in the manufacture of sewage treatment equipment and the advanced technical resources of the latter firm.

Hartleys first supplied equipment for sewage treatment in the early years of this century and is now one of the principal British companies in this field. It offers a complete range of sewage treatment equipment, including scrapers, distributors, sluice gates and digestion plants, much of which has been supplied to the major sewage works in Britain and abroad. Simon-Carves built its first effluent treatment plant more than 10 years ago and has since carried out a large-scale research and development programme. The Biochemical Engineering Division was formed to exploit the results of this work in the industrial effluent field, and Simon-Hartley was founded to apply these results to the field of domestic sewage.

The combination of theory and practice has led to the rapid development of new and better techniques. Specific examples are the Simcar Aerator, the Simcar Respirometer, and the Hartley Heatamix Sludge Digestion System.

The Simcar Aerator has been evolved specifically to provide the high rate of oxygen transfer needed for the more recent activated-sludge processes. Annual sales in the USA by Simon-Hartley's licensees—the Eimco Corporation—are now approximately 2,000,000 dollars.

The Simcar Respirometer was originally developed to improve the accuracy and speed of preparation of the basic data on which the design of effluent treatment plants is based. The results have led to a new and more economical design basis for activated-sludge systems.

The Hartley Heatamix system of sludge digestion ingeniously combines the function of sludge recirculation and heating in one unit, which may be mounted internally or externally to the digester, thus combining the benefits of high velocities at the sludge transfer surfaces and consequent economy in operation.

TUKE & BELL LTD
1 LINCOLN'S INN FIELDS
LONDON WC2, ENGLAND

telephone: 242 8095
telegrams: Autoflush London WC2

For over 50 years Tuke & Bell have been designing small sewage treatment plants to prevent pollution of rivers and streams. During that half-century they have been manufacturing purpose-designed machinery and equipment to fulfil the special functions needed.

Since their inception Tuke & Bell have produced plant of robust and simple construction that will perform under the most arduous conditions with the minimum of attention. For the small sewage treatment plant for a population of up to 300, the firm's 'V' notch method of distribution over filter bed surfaces means months of satisfactory operation without need of attention. Tuke & Bell's larger Carlton distributors for municipal filter beds require no lubrication and, if the distributor arms are fitted with the firm's Shell sprays, cleaning is reduced to a minimum.

Likewise, their pumping equipment is designed to need little maintenance. They particularly recommend their Lift and Force ejectors for flows up to 150 Imp. gallons (0·68 m^3) per minute and their Barrington self-priming pump for volumes up to 1000 gallons (4·50 m^3) per minute. When used with their Brownson floatless control, there are no moving parts in the sumps below ground; all equipment requiring maintenance is housed at ground level, where access is easy; the pump house is properly lighted and ventilated.

Tuke & Bell's latest development is the control of gravity-filled ejectors. There are sites where a below-ground ejector has site advantages. By controlling such plant electromatically these ejectors eliminate entirely all moving parts, such as floats and counterweights, within the ejector body. The compact control panel is housed in an accessible position a short distance away.

Another specially designed Tuke & Bell equipment is the Brownson reciprocal pump, made in three sizes for dealing with sewage tank effluents containing colloidal matter and small solids in suspension, which is capable of discharges down to three gallons (13·6 litres) per minute; together with three larger sizes which will also raise unscreened crude sewage. The largest one, the Twin-Ram Brownson, is designed to raise heavy settlement tank sludge and has an output of 50–80 gallons (227–364 litres) per minute according to the nature of the liquid being pumped.

As specialists in small sewage treatment plants, the firm is able to help in solving the water pollution problems of the smaller communities.

Appendix 2

The following firms are among the most important consulting engineers in Britain and are fully qualified to advise on, design and supervise works for water pollution control, sewerage, and effluent treatment.

W. S. ATKINS & PARTNERS	*telephone: Epsom 26140*
WOODCOTE GROVE, ASHLEY ROAD	*telegrams: Kinsopar*
EPSOM, SURREY, ENGLAND	*telex: Epsom 23497*

Founded in 1938, the company now numbers over 1,200 and is one of the largest consultant organisations in Europe. A specialised group within the firm deals with water supply schemes and effluent treatment as follows:

Water supply: study of requirements and possible sources; planning of schemes for domestic, agricultural and industrial use; control of water quality and design of process plants. Assistance in implementation and seeking possible sources of financial aid. Examples: Legadadi Water Supply Scheme at Addis Ababa, Ethiopia; Abadir Plantation Study, topographic and soil survey with irrigation and land development plan for the Ethiopian Government sponsored by the British Ministry of Overseas Development; Tinsley Park steelworks associated reservoir, distribution and effluent treatment plant for English Steel Corporation Ltd, Sheffield.

Effluent treatment: study of effluent problems on a regional basis and for individual industrial plants; planning of schemes considering the methods available (whether chemical or biological) and economic aspects. Consideration of sludge disposal; purification and re-use of water; disposal of solid wastes. Design of effluent treatment plant, preparation and letting of tenders; supervision of construction; commissioning and hand-over of plant to the client; training of personnel. Examples: an effluent scheme for Henry Wiggin & Co Ltd (producers of nickel alloys) at Hereford, England; sewerage scheme for Drax Power Station for the Central Electricity Generating Board in Yorkshire; water treatment and effluent plant for an integrated pulp and paper mill for the Wiggins Teape Group at Fort William; surveys, plant designs, etc. for a wide range of industries, including ferrous and non-ferrous metallurgical, power supply, food industry (dairy and brewery wastes).

BABTIE SHAW & MORTON
95 BOTHWELL STREET
GLASGOW C2, SCOTLAND

telephone: 248 4211
telegrams: Triumvir Glasgow
telex: Babtie 77202

The firm acts as consulting engineers for a wide range of civil and structural engineering work and has been responsible for water supply installations, hydro-electric projects, drainage, sewage and industrial waste treatment, highways, docks, harbours, shipyards and maritime works, as well as industrial plants including structural engineering in steel, reinforced concrete and pre-stressed concrete.

The firm carries out chemical and biochemical analyses and interpretations as a preliminary to process plant design. It designs pilot-scale plants and, on completion of full-size works, undertakes the training of operating personnel. It also accepts commissions for diagnostic work on existing plants.

It has designed effluent treatment plants for wastes from fermentation, chemicals, fellmongering, textile, tanning, automotive and other industries. Among major projects are: Glasgow (£4,000,000); Greenock, Port Glasgow and Gourock (£4,000,000); Irvine and district (£5,000,000); Perth (£1,000,000); Kirkintilloch (£500,000); Alloa and district (£500,000); Imperial Chemical Industries Ltd (Dyestuffs and Nobel Divisions); The Distillers Co Ltd; Scottish Grain Distillers; Scottish Fellmongers Ltd; National Chrome Tanning Co Ltd; Patons & Baldwins Ltd.

D. BALFOUR & SONS
131 VICTORIA STREET
LONDON SW1, ENGLAND

telephone: 834 1601

The firm, founded in 1882, specialises in designing and supervising the construction of schemes of water supply, sewerage, surface water drainage, sewage and trade waste disposal in Britain and overseas including Arabia, Africa, the eastern Mediterranean and Portugal

It prepares feasibility studies and reports, estimates of capital and operating costs, details of schemes, working drawings, specifications, bills of quantities and contract documents. It gives advice on tenders and supervises the construction of the works, issues certificates for payments to contractors, prepares statements of final costs, record drawings, working instructions, etc. It advises clients on the operation and maintenance of their works and the training of their staff, as well as on the financing of projects, the raising of international loans and the revenue aspect of projects.

Among the schemes with which the firm has been associated are:

	Approx. overall cost
Bybrook sewage disposal works	£2,900,000
Beverley sewage disposal works	£1,325,000
Birtley sewage disposal works	£1,200,000
Eastry sewage disposal works	£1,000,000
Kidderminster purification works	£1,700,000
Snodland purification works	£1,200,000
Normanton Altofts main drainage	£1,000,000
Norwich sewage purification works	£1,875,000
Queenborough-in-Sheppey sewage disposal works	£1,750,000
Reigate sewage disposal works	£1,250,000

BINNIE & PARTNERS
ARTILLERY HOUSE
ARTILLERY ROW
LONDON SW1, ENGLAND

telephone: 799 7050
telegrams: Arbintro London SW1

Binnie & Partners are consulting engineers specialising in sanitary and hydraulic engineering. During the past 10 years they have advised and supervised the construction of sewerage, sewage disposal, drainage, water supply, irrigation and hydro-electric works in many parts of the world, to a total value of over £400,000,000.

The firm's experience in the fields of sewerage, sewage disposal and drainage is of long standing, extending back more than 60 years.

With its wide parallel experience of water supply, the firm is well qualified to advise clients on all aspects of integrated schemes. It is familiar with the requirements of international lending agencies, having been responsible for several schemes financed by the World Bank.

In order to provide a complete service to clients, the firm employs specialists in many fields related to sanitary engineering and has its own laboratory. Among more than 50 projects undertaken in the last 10 years involving disposal and treatment of sewage, including trade effluents, are the following: St. Austell Bay, England—study to prevent pollution of beaches by industrial effluents; Bangalore, India—treatment of 154,000,000 gallons (700,000 m^3) a day of sewage and trade effluents; Spilsby, England—submarine pipeline 1 mile 990 yds (2·5 km) long under construction.

BMMK & PARTNERS (BRIMER, MARTIN, MAGGS, KEEBLE & PARTNERS)
IMBERHORNE LANE
EAST GRINSTEAD, SUSSEX
ENGLAND *telephone: East Grinstead 21277*

The firm was formed in 1962 and has associates in Australia, Honduras, Kenya, Malawi, Nigeria, Rhodesia, the West Indies and Zambia. It has been responsible for a wide range of civil and structural projects, including water supply and irrigation, dams, sewerage and sewage disposal, drainage, land reclamation, municipal refuse composting, township development and services, including roads. It has been consulted by government and local authorities in Ghana, Nigeria, Sierra Leone, Malawi and Zambia, Saudi Arabia, Honduras, and the Caribbean, and maintains permanent offices in a number of these territories.

The partners have been responsible for a number of very economical small sewage purification installations and have adapted the oxidation pond technique to the particular requirements of the clients and areas concerned. They have also acted as special consultants to international bodies such as the World Health Organisation in connection with projects on sewerage, sewage disposal, surface water drainage and municipal refuse disposal for the city of Ibadan in Nigeria.

BYLANDER, WADDELL & PARTNERS
169 WEMBLEY PARK DRIVE
WEMBLEY, MIDDLESEX, ENGLAND *telephone: 904 9511*

In addition to civil and structural engineering in many major commercial and industrial developments, a department was formed in 1960 to deal with the design of effluent treatment plants, acid recovery plants, water pollution and sewage treatment, together with associated piped services for gas and fluids.

The practice is, therefore, able to offer a comprehensive service to industrialists and public authorities, from surveys and feasibility studies to project management, including the design and specification of civil, structural and mechanical engineering, obtainment of tenders, procurement of plant, supervision and commissioning for all aspects of effluent and water treatment work, both in Britain and overseas.

The practice is run by five partners and carries an average of 175 engineers, draughtsmen and administrative personnel. Apart from its head office at Wembley, the firm has branches in Sheffield and Glasgow.

JOHN DOSSOR & ASSOCIATES
WEST HUNTINGTON HALL
YORK, ENGLAND *telephone: York 68102*
and
4 THE MOORS
WORCESTER, ENGLAND *telephone: Worcester 28288*

John Dossor & Associates, consulting civil engineers, was founded in 1944. The partners and senior staff are all chartered engineers with considerable experience in schemes of drainage, sewerage, sewage and trade waste purification and refuse disposal. They undertake responsibility for all stages of project development, from feasibility studies and estimates to outline and detailed design, working drawings, tender documents, supervision of construction and final completion.

The firm and partners are members of the British Consultants Bureau and of the British Association of Consulting Engineers. Their practice is in full accordance with the latter's code of ethics and scales of fees. They have no connection whatever with any manufacturing or contracting organisation. They have been and are responsible for projects worth many millions of pounds, including, for instance, the enlargement of the Borough of Stafford's sewage purification works at a cost of over £700,000.

R. FERGUSON & S. McILVEEN
15 COLLEGE GARDENS
BELFAST BT9 6BU
NORTHERN IRELAND *telephone: Belfast 669461*

Founded in 1922, the firm practises mainly in Northern Ireland and has a staff of about 120. It has been connected with a large amount of civil engineering work throughout the country, and has acted for almost every local authority and government department. It is particularly well known for its work in the field of sewerage and sewage disposal, having designed and supervised the construction of many sewage treatment plants, including some in the traditional form and others of the most modern, fully automated types. It has also undertaken the design and supervision of concrete, rockfill and earthfill dams for impounding reservoirs, many filtration works for the purification of water supply, as well as several thousand miles of pipelines. In addition, it has carried out the design and supervision of major drainage schemes for urban areas, as well as bridge works and classified roads for government and local authorities.

The firm has work in hand on the following sewerage and sewage disposal schemes in Northern Ireland:

Client	Scheme	Approx. overall cost
Newtownards Borough Council	Old town sewerage scheme	£235,000
	Development sewerage and lower drainage schemes	£894,000
Northdown Rural District Council	Killarn sewerage scheme	£70,000
Fermanagh County Council	Enniskillen sewerage and sewage disposal scheme	£530,000
	Tullyhummon sewerage and sewage disposal scheme	£28,000

Omagh Urban District Council	Sewerage and sewage disposal scheme	£440,000
Omagh Rural District Council	Village sewerage schemes	£72,000
Newtownards Joint Sewerage Board	Extension of Ballyrickard sewage disposal works	£325,000
Portadown Borough Council	Sewerage and sewage disposal scheme	£250,000
Ballymena Borough Council	Sewage disposal works	£750,000
South Down Rural District Council	Rostrevor and Murlough, and miscellaneous sewerage	£370,000
Larne Borough Council	High-level sewerage	£81,000
Cookstown Rural District Council	Miscellaneous sewerage	£100,000
Cookstown Urban District Council	Sewage disposal works	£250,000
	Storm sewerage	£150,000
Limavady Rural District Council	Sewage disposal works—Aghanloo	£60,000

SIR ALEXANDER GIBB & PARTNERS
4 TOTHILL STREET
LONDON SW1
ENGLAND

telephone: 01-930 9700
telegrams: Gibbosorum London SW1

Founded in 1922, the firm has been responsible for the design and construction of a wide range of civil engineering and building works at home and abroad, including industrial plants for the treatment and disposal of sewage and effluents from a variety of industrial processes. It has also carried out full-scale investigations for a complete sewage disposal scheme for the city of Tehran, the first stage of which is estimated to cost £5,000,000.

The firm acts as consulting engineers for the Decimal Branch of the Royal Mint at Llantrisant, South Wales. The plant treats the effluent from the pickling works, which contains sulphuric acid tartaric acid and acid wash water as well as heavy metals such as copper, nickel, zinc and iron. The effluent is required to be treated to a high degree of purity, having a maximum content of one milligram per litre of heavy metal and a pH of 7 : 9 before being discharged to the river. Other works for which it has been responsible include effluent treatment plants for the Tilbury 'B' power station and for the Atomic Weapons Testing Range at Maralinga, South Australia

HOWARD HUMPHREYS & SONS
8 FRANCIS STREET — *telephone: 834 5875*
LONDON SW1, ENGLAND — *telegrams: Cutaneal Sowest London*

The firm employs some 460 qualified engineers and has its head office in London and design offices at Epsom, Reading and Cardiff It is represented through its own branch offices and its association with other firms in the Bahamas, Bermuda, Ceylon, Ghana, Honduras, Ireland, Italy, Jamaica, Kenya, Lebanon, Libya, Malawi Nigeria, Uganda and Zambia.

The practice covers a wide field of civil engineering in many parts of the world and provides a complete range of consulting engineering services. In addition to dealing in particular with all forms of sewerage, surface water drainage and sewage purification, the firm also specialises in dams and all aspects of water supply, highways bridges and tunnels, gas pipelines and power stations, structures foundations, irrigation and coastal works.

Among the more important sewerage schemes in hand or recently completed are those for various towns in Libya: Benghazi (£11,500,000), Derna (£2,400,000), Misurata (£2,400,000), Sebha (£650,000), Tobruk (£1,850,000), Tripoli (£11,500,000); for Khartoum, Sudan (£3,000,000); Kumasi, Ghana (£3,500,000) Kampala, Uganda (£1,500,000); Dar-es-Salaam, Tanzania (£1,450,000); Nairobi, Kenya (£1,062,000); Famagusta, Cyprus (£1,100,000); Bermuda (£500,000); and Colombo, Ceylon (£3,000,000).

HUSBAND & CO
388 GLOSSOP ROAD
SHEFFIELD S10 2JB
ENGLAND
and at
ST ERMIN'S, CAXTON STREET
LONDON SW1, ENGLAND

telephone: Sheffield 28834
telegrams: Consulting Sheffield

telephone: 222 1917
telegrams: Husbanco London

The firm has its head office in Sheffield and a London office, each having a staff of over 100 qualified engineers, as well as branches in Greece, Persia and Ceylon. Husband & Co has been retained as consultant for the design layout, co-ordination and supervision of construction of large effluent treatment plants in Britain and overseas. An extensive department designs schemes for domestic and industrial water supply, sewerage and sewage disposal. From an analysis of the trade waste, the department is able to design and specify plant capable of producing an effluent to meet the requirements of the appropriate sewerage or river authority. Economic assessments are made of the capital and running costs of alternative methods of treatment.

Among the more important sewage treatment plants the firm has either recently completed or has in hand are: works in Britain valued at £4,000,000; report and basic design of drainage and sewage purification for Bangkok, Thailand (£12,000,000); report and basic design of drainage of towns south of Colombo, Ceylon (£15,000,000); and reports on drainage of Anuradhapura (£500,000) and Peradeniya University (£500,000), both in Ceylon.

G. B. KERSHAW & KAUFMAN
CHANDOS HOUSE, PALMER STREET
LONDON SW1, ENGLAND

telephone: 222 2111
telegrams: Outfall London SW1

The firm specialises in the design and execution of schemes of sewerage, sewage disposal, trade waste treatment and water supply. Work at present in hand in Britain includes sewers up to eight feet (2·4 metres) in diameter, the centralisation of sewage treatment works, advice on industrial wastes, and water supply projects aggregating £12,000,000 in value, while work abroad has been of the order of £10,000,000.

Among the schemes with which the partnership has been associatec overseas are a £2,500,000 sewerage and sewage disposal plan comprising trunk sewers, sea outfall and comminutor bay and a feasibility investigation for a £7,500,000 project of the same kind both in the Middle East.

The founder of the practice, the late G. B. Kershaw, was pre-viously engineer to the Royal Commission on Sewage Disposal The present senior partner, H. P. Kaufman, represents the Associa-tion of Consulting Engineers on the British Standards Institution's Code of Practice Sewerage Committee, is chairman of the panel deal-ing with design and construction and is a member of the Ministry of Housing and Local Government's working party on the design anc construction of underground pipelines.

ERIC G. LEDIARD
1 RIDLEY PLACE
NEWCASTLE UPON TYNE NE1 8JQ
ENGLAND *telephone: Newcastle 20647*

The practice was established in 1946 and is now almost exclusively engaged on schemes of main drainage, sewerage and sewage disposal and treatment of industrial wastes. Among the many industria estate development schemes for which the firm has been responsibl on behalf of the English Industrial Estates Management Corpora-tion, the Teesside development alone involved contracts totalling £1,000,000 for the survey of the site, design and supervision of the earthworks, main drainage and estate roads.

Its urban development schemes include one for Thornaby-on-Tees, for which it prepared the plan and development report, being subsequently retained to design and supervise the construction of the main site and off-site drainage works and the main roads. Similar work of main drainage, highway design and construction including bridge works was undertaken for the Billingham Urban Distric Council for its new town at Roseberry Road.

In the early days of the practice, many surveys were undertaken for the Forestry Commission in connection with new villages in the Kielder Forest area of Northumberland and in the design anc supervision of water supply, sewerage and sewage disposal schemes Currently, schemes of main drainage, flood relief works, sewage treatment by the activated-sludge and biological filtration processes are under construction or in the design stage. Feasibility reports are in preparation for regional projects of sewerage and sewage disposal.

MCLELLAN AND PARTNERS
SHEER HOUSE
WEST BYFLEET, SURREY
ENGLAND

telephone: Byfleet 43271
telegrams: Mclelan Byfleet
telex: 262340

McLellan and Partners have been consultants for industrial waste and effluent schemes both in Britain and overseas. Recent and current work includes: treatment for re-use or disposal of 2,000,000 gallons (9,100 m^3) per day of waste effluents from the Highveld Steel & Vanadium Corporation's integrated iron and steel works at Witbank, South Africa; treatment for re-use of 750,000 gallons (3,400 m^3) per day of contaminated circulating water in the integrated iron and steel works of New Zealand Steel Ltd, Auckland, New Zealand; neutralisation of acid and alkali effluents from industrial processes; reports on treatment and re-use of river water for ceramic manufacture and on the recovery of chromic acid from chrome swill.

McLellan and Partners have extensive experience of refuse disposal, and this enables them to advise on the most economical method of disposal of solids from treated liquid effluent, including incineration with or without waste heat boilers, pre-treatment and controlled tipping or composting.

MERZ & MCLELLAN
AMBERLEY, KILLINGWORTH
NEWCASTLE UPON TYNE NE12 ORS
ENGLAND

telephone: Newcastle upon Tyne 663955
telegrams: Amber Newcastle-on-Tyne
telex: 53561

Founded in 1899, the firm is a partnership of professional engineers entirely independent of any manufacturing interests. It employs a total staff of over 1,100, more than 600 of whom are chartered engineers and technicians. Besides its Newcastle main office, it has offices in London, Esher, Edinburgh, Sydney, Melbourne and Perth (Western Australia), with associated firms in South Africa and Rhodesia. Currently, work is in hand in Singapore, South America, Iran and West Africa.

Its main fields of activity are in public electrical power supply and heavy industry. Effluent treatment projects dealt with by the firm are often associated with these major activities, but it undertakes studies, design and management of effluent projects independently. Recent works include: delay ponds for irradiated fuel elements and rehabilitation of sedimentation and biological filter for domestic sewage at the Chapelcross plutonium works of the United Kingdom Atomic Energy Authority; extended aeration and biological filtration for domestic sewage and neutralisation of waste from water treatment plant at West Burton, Nottinghamshire; millscale recovery pits for steelworks at Newcastle (New South Wales) and Whyalla (South Australia); chlorination for control of marine growths in sea outfalls at Blyth (Northumberland), Dampier (Western Australia), Longannet (Scotland), Aberthaw (Wales) and Jurong (Singapore)

PARSONS, BROWN & PARTNERS
11 WATERLOO PLACE
LONDON SW1, ENGLAND

telephone: 930 7781 and 0881
telegrams: Parbro London
telex: 916549

The firm has a technical staff of 150 and has executed major work in town planning and civic centres, including sewerage and water supply schemes; industrial complexes and trading estates, including the treatment of trade and domestic effluent; and many large buildings in which domestic, storm and chemical waste drainage schemes have been incorporated. The firm has also been responsible for the design and supervision of effluent pipelines on land and in the sea.

Many of the firm's engineers have considerable experience in sewage treatment and water supply, both in Britain and overseas including tropical countries. In the field of sewage treatment, th firm undertakes the design of sewers, storm water drains, pumping stations, pipelines and sewage treatment plants. It has specialised experience in the design of low-cost oxidation lagoons and of mor sophisticated, highly mechanised installations.

Among the projects for which the firm has been responsible is feasibility study and master plan for the city of Peterborough covering foul and storm drainage systems as well as the selectio of site and preliminary design for a new sewage treatment plant Estimated cost of works £6,500,000.

PICK, EVERARD, KEAY & GIMSON
6 MILLSTONE LANE
LEICESTER LE1 5JD
ENGLAND *telephone: Leicester 25341*

Established over a century ago, the firm has specialised in public health engineering as well as in industrial and commercial development and has considerable experience in the treatment of sewage and trade effluent. It has been consulted by numerous public authorities, including government departments, hospital boards and water boards, not only about treatment processes, but also on the design and supervision of construction of the structures associated with these processes.

The firm is at present advising over 50 public authorities in England in matters of sewage treatment, involving in many cases the treatment of trade effluent. Among the many projects with which it has been associated, the following are examples of those undertaken for public authorities in progress or completed within the last three years: Ashby-de-la-Zouch (overall cost: £350,000); Barrow-upon-Soar (£430,000); Grantham (£770,000); Wigston (£325,000); Witham (£350,000).

W. H. RADFORD & SON
26 ADDISON STREET
NOTTINGHAM NG1 4GY *telephone: Nottingham 45451 and 47765*
ENGLAND

Since 1885, W. H. Radford & Son have advised 97 local authorities on the treatment and disposal of sewage and trade effluents. (They have also advised many of those and others on sewerage and water supply.) During the past 10 years they have been engaged on the following sewage and/or effluent treatment projects, some of which were or are extensions of schemes previously carried out under the firm's direction: Alsager (£327,000); Biggleswade (£306,000); Blandford Forum (£29,000); Denbigh (£90,000); Felling (£63,000); Kirkby-in-Ashfield (£221,000); Louth (£299,000); March (£99,000); Matlock (£175,000); Melton Mowbray (£102,000); Skegness (£74,000); Thorne (£365,000); Thurrock (£1,950,000).

RESOURCES GROUP
CHANDOS COURT
CAXTON STREET
LONDON SW1, ENGLAND

telephone: 222 2623
telegrams: Resources London SW1

The Resources Group, comprising Rofe & Raffety, C. H. Dobbie & Partners, Livesey & Henderson, and Sandford, Fawcett, Wilton & Bell, is a partnership of consulting engineers whose experience goes back more than 100 years and covers a wide field. In recent years the Group has become increasingly engaged in projects overseas, notably in the Lebanon, in Uganda (sponsored by the World Health Organisation), and in Saudi Arabia (the cost of which is expected to run into some £20,000,000).

Public health engineering has always been an important feature of the Resources Group's activities. Sewerage schemes valued at more than £20,000,000 are now being undertaken by the Group in Britain, including trunk sewerage schemes, treatment works, and the design of one of the largest sea outfalls yet planned. Similarly, the Pasveer ditch at Scunthorpe is one of the largest constructed in Britain to date. The Group employs chemists for advice on sewage treatment works, lagoons, etc. When undertaking work for its clients, the Group is pleased to give training facilities in the specialist techniques to nominated members of their staffs.

The Group is at the disposal of prospective clients wishing to inspect its current works. These include:

Overseas	Overall cost
Uganda: Master plan for water supply and sewerage, Greater Kampala and Jinja	—
Lebanon: Commissioning of sewerage scheme, Jounieh	£150,000
Britain	
Aethwy, Anglesey: Sewerage and treatment works	£750,000
Braintree and Bocking, Essex: Sludge treatment plant	£500,000
Canterbury, Kent: Sludge treatment plant	£450,000
Dunstable, Bedfordshire: Sewage treatment works	£250,000
Ennerdale, Cumberland: Sewerage and treatment works	£600,000
Honiton, Devon: Industrial effluent treatment plant	£169,000
Melford, Suffolk: Effluent treatment, lagoons, etc.	£755,000

Nantwich, Cheshire: Pumping stations and treatment works	£585,000
Nantwich, Cheshire: Model analysis to investigate effluent separation	(at research stage)
Rhymney Valley, South Wales: Sewerage system, tanks and outfall works	£1,530,000
Rochford, Essex: Effluent treatment works	£1,200,000
Scunthorpe, Lincolnshire: Sludge treatment plant and Pasveer ditch	£1,500,000
Western Valleys, Monmouthshire: Sewerage system and sea outfalls	£2,900,000
Twrcelyn, Anglesey: Sewerage and treatment works	£600,000
Wrexham, Denbighshire: Sewage disposal works	£2,000,000

SCOTT WILSON KIRKPATRICK & PARTNERS
WINSLEY STREET
LONDON WIN 7AQ
ENGLAND

telephone: 580 6688
telegrams: Pontifact London W1
telex: 21724

Scott Wilson Kirkpatrick & Partners was formed in 1954 by the malgamation of Sir Cyril Kirkpatrick & Partners and Scott & Wilson. Associated firms of the same name, each operating under he direct control of a resident partner, have been established in Africa and Hong Kong. These partnerships now embrace 12 partners nd 10 associates. Staff numbers over 500 throughout the world and ncludes 44 key personnel, another 250 qualified engineers and some 20 technicians.

It is the policy of the firm to maintain close contact with research nd to encourage the use of new methods of design and construction. n the field of sanitary engineering, the firm has specialised in the se of controlled discharge to sea incorporating long outfalls of lastics pipe.

Among the projects with which the firm has been associated are he following:

Country	Scheme and progress	Approx. overall cost
United Kingdom	Hunstanton sewage disposal scheme (completed 1965)	£120,000

United Kingdom	Sandwich sewage disposal (construction due to start in 1969)	£450,000
United Kingdom	Prestatyn foul sewerage (construction due to start in 1969)	£500,000
United Kingdom	Unigate Ltd, Burnham, disposal of trade effluent (design in hand)	£40,000
United Kingdom	Gravesend sewage treatment works (submitted January 1969)	Report
Nigeria	Kaduna Textile Ltd, Kaduna, sewage disposal (completed 1967)	£25,000
British Honduras	New capital sewage treatment works (due to be completed in 1970)	£250,000

SIR FREDERICK SNOW & PARTNERS
144 SOUTHWARK STREET
LONDON SE1, ENGLAND
telephone: 928 5688
telegrams: Efsnowpar London

Founded in 1943, this firm has undertaken a number of importar sewerage schemes, including the design of treatment plants for publi and local authorities in Britain and overseas, as well as for variou industrial undertakings. The schemes have included both th percolating filter and the aeration systems of treatment. It ha prepared a series of reports for the Greater London Council in whic the possibility of transporting the digested sludge from the Londo sewage treatment works to sea by pipeline has been investigated. Th reports have dealt with the laying of 100 miles (161 kilometres) c twin pipelines on land and approximately nine miles (14·5 kilo metres) of a twin submarine outfall, together with ancillary faciliti and pumping stations.

The firm also undertakes surface water drainage, the most recer scheme being one for the London Borough of Haringey, where modified version of the Road Research Laboratory's hydrograp method was used. The first phase of the recommended scheme is t be implemented in 1969. Other surface water investigations hav been carried out for Barcelona airport and for industry.

The partnership has designed and constructed water supplies fo the University of Warwick and for a new town in Trinidad, as we as for various industrial clients.

Additional schemes in progress include: Rochford/Great Wakering sewage scheme (overall cost: £225,000); London Borough of Southwark—relaying Park Street sewer (£37,000); ICI Blackley—culverting of river Irk (£150,000).

JOHN TAYLOR & SONS
ARTILLERY HOUSE, ARTILLERY ROW
LONDON SW1, ENGLAND

telephone: 222 7050
t legrams: Waterage London SW1

Founded in 1869, the firm specialises in public health engineering and deals with all aspects of schemes for water supply and treatment, drainage, sewerage, and sewage and trade effluent treatment. There are at present six partners, two being direct descendants of the founder.

The largest part of the practice is in Britain, but it has recently carried out work in many countries overseas, as indicated by the list below.

The firm has advised many municipal authorities in Britain on effluent treatment and has recently been responsible for the construction of sewage treatment works for the cities of Gloucester and Worcester. It is currently engaged on the design of a new sewage treatment plant for West Hertfordshire, as well as on many other similar projects. It specialises in schemes of sewage disposal to sea. It is acknowledged as one of the leading authorities on the treatment of industrial effluents.

Among overseas schemes at present in hand, the following are of particular interest:

	Approx. overall cost
Iraq: Sewers and sewage treatment works for Baghdad	£30,000,000
Aden: Sewers and sea outfall	£2,000,000
Jamaica: Sewers and sewage treatment works for Kingston and Montego Bay	£2,200,000
Eire: Extension to sewerage system and sewage treatment works in Dublin	£2,000,000
Qatar: Sewage treatment works for Doha	£1,500,000

WARD, ASHCROFT & PARKMAN
CUNARD BUILDING
LIVERPOOL L3 1ES
ENGLAND

telephone: 236 6363
telegrams: Pipeline Liverpool 3
cablegrams: Pipeline Liverpool 3
telex: WAP 627110 Chamcom Lpool

Ward, Ashcroft & Parkman, founded in 1888, provides a comprehensive service covering all aspects of civil and structural engineering, including surveys and feasibility reports, the design and preparation of contracts, estimates, quantity surveying and supervision of construction. The firm has a mechanical engineering department.

Since its inception, the firm has been responsible for many large schemes, including the Runcorn transporter bridge, the Tranmere Bay Development Company's dock system, now known as Cammell Lairds, the water supply for Milford Haven and main drainage for the Amman Valley Joint Sewerage Board. In 1956-59, the firm carried out a detailed study of the tidal barrage scheme at Milford Haven; this being the first such scheme to be taken through Parliament in this country.

WATERHOUSE & PARTNERS
NEWCASTLE HOUSE, HIGH SPEN
ROWLANDS GILL, CO. DURHAM

telephone: Rowlands Gill 2251
telegrams: Consult Rowlandsgill

The firm was founded in 1945 and specialises in water supply, sewerage and sewage disposal, as well as in marine and dock work and the construction of reinforced concrete and steel structures. It has carried out many sewerage schemes for local authorities, ranging from small village schemes to large urban districts and boroughs.

Work is in progress on the sewerage for the new town of Cramlington, in north-east England, at an overall cost of £5,000,000; for the Mid-Tyne sewerage scheme; and for large private developments. Other projects in hand include extensions to the Browney sewage works (including milk wastes) costing £350,000; extensions to the Fareham-Salterne Lane works and abattoir wastes (£1,000,000); and the Londonderry main drainage works (£1,600,000).

A. H. S. WATERS & PARTNERS
85 NEWHALL STREET
BIRMINGHAM 3, ENGLAND *telephone: 236 0493*

The practice was founded by A. H. S. Waters in 1919 and became a partnership in 1955. It specialises in sewerage, sewage disposal plants, treatment works for trade effluents, water supply and structural engineering.

It has been consulted and is retained by numerous county councils, local authorities, water authorities and river authorities. The partners have acted as expert witnesses on many occasions before parliamentary committees and other tribunals.

The value of work currently in hand is in excess of £50,000,000.

The major projects for which the firm is responsible include at present: storm water drainage, main sewerage and sewage disposal works for development towns and others, including Crewe, Daventry, Droitwich and Tamworth, at costs of up to £2,500,000 each; regional works of sewerage and sewage disposal for several large urban and rural areas, at costs of up to £1,500,000 each; regional and other works of water supply for various water boards, including Cardiganshire, Herefordshire, Radnorshire and North Breconshire, at costs of up to £1,250,000 each.

. D. & D. M. WATSON
67 TUFTON STREET *telephone: 222 1564*
LONDON SW1, ENGLAND *telegrams: Culvert London SW1*

The firm was founded in 1909 and specialises in the supply of water and the treatment and disposal of waste waters. It has wide experience in the assessment of water resources, water treatment and distribution, land drainage, foul sewerage, sewage treatment, and river, estuarine and marine pollution surveys, in the United Kingdom and in arid and tropical areas, including the Arabian Gulf and South-East Asia. The firm has a branch office in Singapore and there is an associate firm in Saudi Arabia.

Some of the larger projects include:

Client	Scheme	Approx. value
G.L.C./Middlesex C.C.	East and West Middlesex main drainage and sewage disposal	£20,000,000

Teesside County Borough	Investigation and Report on sewerage and sewage disposal	£15,000,000
Milton Keynes D.C.	Investigation and Report on surface water drainage and sewerage, and sewage treatment design	£26,000,000
Harlow & Stevenage D.C.	Middle Lee Regional sewage treatment works	£12,000,000
Tyneside Joint Sewerage Board	Marine investigations and design of sewerage and sewage treatment works	£10,000,000
Derby County Borough	Sewage treatment works	£5,000,000
City of Leicester	Sewage treatment works	£3,000,000
Brunei Government	Sewerage and sewage treatment	£1,000,000
Dubai Government	Sewerage and sewage treatment	£2,000,000
Hong Kong Government	Reports on sewerage and sewage disposal and marine investigation	£16,000,000
Kuala Lumpur Municipality	Sewerage and sewage treatment	£3,000,000
Saudi Arabian Government	Reports and design of surface water drainage, sewerage, sewage treatment and refuse disposal, in Mecca, Jeddah and other towns	£39,000,000
Singapore Government	Sewerage, sewage treatment and marine investigations	£10,000,000

Appendix 3

Contact points for manufacturers and consultants

BRITISH SEWAGE PLANT MANUFACTURERS ASSOCIATION
9 CADOGAN STREET
LONDON SW3, ENGLAND *telephone: 589 4991*

Great Britain has the world's fourth highest population density but no broad fast-flowing rivers. The treatment of sewage and trade effluent in this country must therefore be as perfect as human intelligence, technical experience and engineering skill can devise.

British sewage plant manufacturers, backed by first-class research and development facilities and over a century of experience, provide the full range of equipment required in modern installations. The member companies of their Association, listed below, produce a very large proportion of the total output in Britain. They welcome inquiries, which may be sent either directly or through the Association.

MEMBER COMPANIES

ACTIVATED SLUDGE LIMITED

AMES CROSTA MILLS & COMPANY LIMITED

DORR-OLIVER COMPANY LIMITED

WILLIAM E. FARRER LIMITED

HAM, BAKER & COMPANY LIMITED

NORSTEL & TEMPLEWOOD HAWSKLEY LIMITED

SIMON-HARTLEY LIMITED

TUKE & BELL LIMITED

WHITEHEAD & POOLE LIMITED

THE ASSOCIATION OF CONSULTING ENGINEERS
2 VICTORIA STREET
LONDON SW1, ENGLAND

telephone: 222 6557
telegrams: Conseng London SW1

The Association of Consulting Engineers was formed in 1913 to regulate the profession, promote its advancement and voice its collective views. It is a voluntary, non-statutory body, yet the strict rules of professional conduct it has enacted are observed not only by its members, but generally by British consulting engineers. All major client organisations now expect and require abidance by these rules.

The qualifications for membership are equally stringent and include fellowship or membership of one or more of the senior professional engineering institutions, as well as substantial experience in practice. A special committee scrutinises every application.

The code of conduct to which all members subscribe ensures the complete independence of their professional judgment by prohibiting them to have any contracting or manufacturing interest: they work solely for their client, from whom alone they receive remuneration. They are forbidden to advertise or solicit for work.

The Association is thus able to recommend its members in full confidence as competent to advise clients in the appropriate engineering specialisation. It points with legitimate pride to the steadily increasing capital value of the work done by its members for overseas clients, which now stands at well over £1,600,000,000.

BRITISH CONSULTANTS BUREAU
55–58 PALL MALL
LONDON SW1, ENGLAND

telephone: 839 7687
telegrams: Britburo London SW1
telex: 916843

The British Consultants Bureau is an organisation of consulting firms working abroad or keen to do so. Its membership includes firms of consulting engineers of various disciplines, planners, surveyors, architects, economists and management experts. Its object is to put client and consultant in touch with each other. A client looking for a particular service will be introduced to a specialist consultant.

BCB is supported by the British Government as well as by its member firms. It is non-profit-making and therefore charges no fees and accepts no commission for its services. Many of its engineer nembers specialise in matters relating to environmental health, uch as water supply, main drainage, refuse disposal, and the revention and control of water pollution. Such firms have a wealth f experience in all kinds of climate and terrain, and they take care o keep in touch with the latest technical developments.

Because of the strict standards of water pollution control required y British legislation, and the need to have complete mastery of the ountry's water resources in such a densely populated country, British know-how in this important subject is rated very highly. irms of consulting engineers specialising in this subject are happy o collaborate with firms in other countries, or to work direct for a lient. In either case, BCB will be glad to bring the two parties ogether.

Printed in England by trade union labour for
Her Majesty's Stationery Office by
Headley Brothers Ltd, Invicta Press, Ashford, Kent